PHILIP'S

**BRITAIN
& IRELAND**

GEL
NBEST

STARGAZING
2022

MONTH-BY-MONTH GUIDE TO THE NIGHT SKY

www.philips-maps.co.uk

Published in Great Britain in 2021 by Philip's,
a division of Octopus Publishing Group Limited
(www.octopusbooks.co.uk)
Carmelite House, 50 Victoria Embankment,
London EC4Y 0DZ
An Hachette UK Company (www.hachette.co.uk)

TEXT
Nigel Henbest © 2021 pp. 4–85
CPRE © 2017 pp. 90–91
Philip's © 2021 pp. 1–3
Robin Scagell © 2021 pp. 86–89

MAPS
pp. 92–95 © OpenStreetMap contributors, Earth
Observation Group, NOAA National Geophysical Data
Center. Developed by CPRE and LUC.

ARTWORKS © Philip's

ISBN 978-1-84907-587-9

Printed in China

Cover: Helix Nebula
Title page: Comet NEOWISE.

CONTENTS

Welcome to the latest edition of *Stargazing*! Within these pages, you'll find your complete guide to everything that's happening in the night sky throughout 2022 – whether you're a beginner or an experienced astronomer.

With the 12 monthly star charts, you can find your way around the sky on any night in the year. Impress your friends by identifying celestial sights ranging from the brightest planets to some pretty obscure constellations.

Every page of the *Stargazing 2022* guide is bang up-to-date, bringing you everything that's new this year, from shooting stars to eclipses. And I'll start with a run-down of the most exciting sky sights on view in 2022 (opposite).

THE MONTHLY CHARTS

A reliable map is just as essential for exploring the heavens as it is for visiting a foreign country. For each month, I provide a circular **star chart** showing the whole evening sky.

To keep the maps uncluttered, I've plotted about 200 of the brighter stars (down to third magnitude), which means you can pick out the main star patterns – the constellations. (If I'd shown every star visible on a really dark night, there'd be around 3000 stars on each chart!) I also show the ecliptic: the apparent path of the Sun in the sky, it's closely followed by the Moon and planets as well.

You can use these charts throughout the UK and Ireland, along with most of Europe, North America and northern Asia – between 40 and 60 degrees north – though the detailed timings apply specifically to the UK and Ireland.

USING THE STAR CHARTS

It's pretty easy to use the charts. Start by working out roughly your compass points. South is where the Sun is highest in the sky during the day; east is roughly where the Sun rises, and west where it sets. At night, you can find north by locating the Pole Star – Polaris – by using the stars of the Plough (see December's Constellation).

The left-hand chart then shows your view to the north. Most of these stars are visible all year: the circumpolar constellations wheel around Polaris as the seasons progress. Your view to the south appears in the right-hand chart; it changes much more as the Earth orbits the Sun. Leo's prominent 'sickle' is high in the spring skies. Summer is dominated by the bright trio of Vega, Deneb and Altair. Autumn's familiar marker is the Square of Pegasus; while the stars of Orion rule the winter sky.

During the night, our perspective on the sky alters as the Earth spins around, making the stars and planets appear to rise in the east and set in the west. The charts depict the sky in the late evening (the exact times are noted in the captions). As a rule of thumb, if you are observing two hours later, then the following month's map will be a better guide to the stars on view – though beware: the Moon and planets won't be in the right place!

THE PLANETS, MOON AND SPECIAL EVENTS

The charts also highlight the **planets** above the horizon in the late evening.

HIGHLIGHTS OF THE YEAR

- **Night of 3/4 January:** maximum of the Quadrantid meteor shower. With the Moon out of the way, this year's display should be spectacular.
- **Night of 12/13 January:** the Moon passes half a degree above dwarf planet Ceres.
- **26 January:** the Moon moves in front of the double star Zubenelgenubi in Libra.
- **3 February:** Jupiter lies to the lower right of the crescent Moon.
- **5 April, am:** Mars passes below Saturn, to the right of brilliant Venus.
- **Night of 21/22 April:** it's an excellent year for observing the maximum of the Lyrid meteor shower.
- **1 May, am:** the two brightest planets, Venus and Jupiter, are very close together.
- **13 May:** the Moon moves in front of Porrima in Virgo.
- **16 May:** an eclipse of the Moon is total in the Americas and parts of Africa and Europe. From the British Isles, the eclipsed Moon is low in the south-west in the early morning.
- **29 May, am:** Mars passes just below Jupiter.
- **1 June, am:** Comet Schwassman-Wachman-3 may treat us to a spectacular storm of shooting stars, the Tau Herculid meteor shower.
- **22 June, am:** the Moon lies between Jupiter and Mars.
- **26 June, am:** Venus and the crescent Moon make a stunning sight.
- **13 July:** the nearest and brightest supermoon of 2022.
- **19 July, am:** Jupiter lies just above the Moon.
- **6 August:** the Moon moves in front of Dschubba in Scorpius.
- **Night of 12/13 August:** maximum of the Perseid meteor shower. Sadly, it's washed out by the light of the Full Moon.

- **14 August:** Saturn is opposite to the Sun in the sky, and closest to the Earth.
- **14 September:** the Moon moves in front of Uranus.
- **16 September:** Neptune is opposite to the Sun, and at its nearest to the Earth.
- **26 September:** Jupiter is at its closest to the Earth and opposite to the Sun in the sky.
- **Night of 21/22 October:** it's a great year for observing the Orionid meteor shower, until the Moon rises around 3.30 am.
- **25 October:** a partial solar eclipse is visible from north-east Africa, the Middle East and Europe. For the British Isles, about 15 per cent of the Sun is covered at maximum (a few minutes before 11 am).
- **30 October to 30 November:** the Taurid meteor shower may treat us to an unusual display of brilliant 'Halloween fireballs'.
- **8 November:** a total eclipse of the Moon is visible from North America and the Pacific, but not from the British Isles.
- **9 November:** Uranus is opposite to the Sun in the sky and at its closest to Earth.
- **Night of 17/18 November:** maximum of the Leonid meteor shower; view them before the Moon rises at midnight.
- **1 December:** Mars is nearer to the Earth than it has been since October 2020.
- **5 December:** the Moon moves in front of Uranus.
- **8 December, am:** Mars suffers a rare occultation by the Full Moon, just minutes after reaching opposition.
- **Night of 13/14 December:** look out for the bright slow shooting stars of the Geminid meteor shower, though this year the show is spoilt by the Moon rising around 9 pm.

I've indicated the track of any comets known at the time of writing; though I'm afraid I can't guide you to a comet discovered after the book has been printed!

I've plotted the position of the Full Moon each month, and also the **Moon's position** at three-day intervals before and afterwards. If there's a **meteor shower** in the month, I mark its radiant – the position from which the meteors stream.

The **Calendar** provides a daily guide to the Moon's phases and other celestial happenings. I've detailed the most interesting in the **Special Events** section, including close pairings of the planets, times of the equinoxes and solstices and – most exciting – **eclipses** of the Moon and Sun.

Check out the **Planet Watch** page for more about the other worlds of the Solar System, including their antics at times they're not on the main monthly charts. I've illustrated unusual planetary and lunar goings-on in the **Planet Event Charts**. And there's a full explanation of all these events in **Solar System 2022** on pages 80–82.

MONTHLY OBJECTS, TOPICS AND PICTURES

Each month, I examine one particularly interesting **object**: a planet, a star or a galaxy. I also feature a spectacular **picture** – taken by an amateur based in Britain – describing how the image was captured, and subsequently processed to enhance the end result. And I explore a fascinating and often newsworthy **topic**, ranging from exotic star names to the centre of the Galaxy.

GETTING IN DEEP

There's a practical **observing tip** each month, helping you to explore the sky with the naked eye, binoculars or a telescope.

Check out the guide to the **Top 20 Sky Sights**, such as nebulae, star clusters and galaxies. You'll find it on pages 83–85.

It's followed by equipment expert **Robin Scagell**'s behind-the-scenes look at how the contributing photographers took the amazing pictures that adorn this book.

If you're plagued by light pollution, use the **dark-sky maps** (see pages 90–95). They show you where to find the blackest skies in Great Britain, and enjoy the most breathtaking views of the heavens.

So: fingers crossed for good weather, beautiful planets, a multitude of meteors and – the occasional surprise.

Happy stargazing!

JARGON BUSTER

Have you ever wondered how astronomers describe the brightness of the stars or how far apart they appear in the sky? Not to mention how we can measure the distances to the stars? If so, you can quickly find yourself mired in some arcane astro-speak – magnitudes, arcminutes, light years and the like.

Here's our quick and easy guide to busting that jargon:

Magnitudes
It only takes a glance at the sky to see that some stars are pretty brilliant, while many more are dim. But how do we describe to other people how bright a star appears?

Around 2000 years ago, ancient Greek astronomers ranked the stars into six classes, or **magnitudes**, depending on their brightness. The most brilliant stars were first

magnitude, and the faintest stars you can see came in at sixth magnitude. So the stars of the Plough, for instance, are second magnitude while the individual Seven Sisters in the Pleiades are fourth magnitude.

Mars (magnitude –1.6 here) shines a hundred times brighter than the Seven Sisters in the Pleiades, which are around 5 magnitudes fainter.

Today, scientists can measure the light from the stars with amazing accuracy. (Mathematically speaking, a difference of five magnitudes represents a difference in brightness of one hundred times.) So the Pole Star is magnitude +2.0, while Rigel is magnitude +0.1. Because we've inherited the ancient ranking system, the brightest stars have the *smallest* magnitude. In fact, the most brilliant stars come in with a negative magnitude, including Sirius (magnitude –1.5).

And we can use the magnitude system to describe the brightness of other objects in the sky, such as stunning Venus, which can be almost as brilliant as magnitude –5. The Full Moon and the Sun have whopping negative magnitudes!

At the other end of the scale, stars, nebulae and galaxies with a magnitude fainter than +6.5 are too dim to be seen by the naked eye. Using ever larger telescopes – or by observing from above Earth's atmosphere – you can perceive fainter and fainter objects. The most distant galaxies visible to the Hubble Space Telescope are ten billion times fainter than the naked-eye limit.

Here's a guide to the magnitude of some interesting objects:

Sun	–26.7
Full Moon	–12.5
Venus (at its brightest)	–4.7
Sirius	–1.5
Betelgeuse	+0.4
Polaris (Pole Star)	+2.0
Faintest star visible to the naked eye	+6.5
Faintest star visible to the Hubble Space Telescope	+31

Degrees of separation

Astronomers measure the distance between objects in the sky in **degrees** (symbol °): all around the horizon is 360°, while it's 90° from the horizon to the point directly overhead (the zenith).

As we show in the photograph, you can use your hand – held at arm's length – to give a rough idea of angular distances in the sky.

For objects that are very close together – like many double stars – we divide the degree into 60 arcminutes (symbol '). And for celestial objects that are very tiny – such as the discs of the planets – we split each arcminute into 60 arcseconds (symbol "). To give you an idea of how small these units are, it takes 3600 arcseconds to make up one degree.

Here are some typical separations and sizes in the sky:

Length of the Plough	25°
Width of Orion's Belt	3°
Diameter of the Moon	31'
Separation of Mizar and Alcor	12'
Diameter of Jupiter	45"
Separation of Albireo A and B	35"

How far's that star?

Everything we see in the heavens lies a long way off. We can give distances to the planets in millions of kilometres. But the stars are so distant that even the nearest, Proxima Centauri, lies some 40 million million kilometres away. To turn those distances into something more manageable, astronomers use a larger unit: one **light year** is the distance that light travels in a year.

One light year is about 9.46 million million kilometres. That makes Proxima Centauri a much more manageable 4.2 light years away from us. Here are the distances to some other familiar astronomical objects, in light years:

Sirius	8.6
Polaris	440
Centre of the Milky Way	26,700
Andromeda Galaxy	2.5 million
Most distant galaxies seen by the Hubble Space Telescope	13 billion

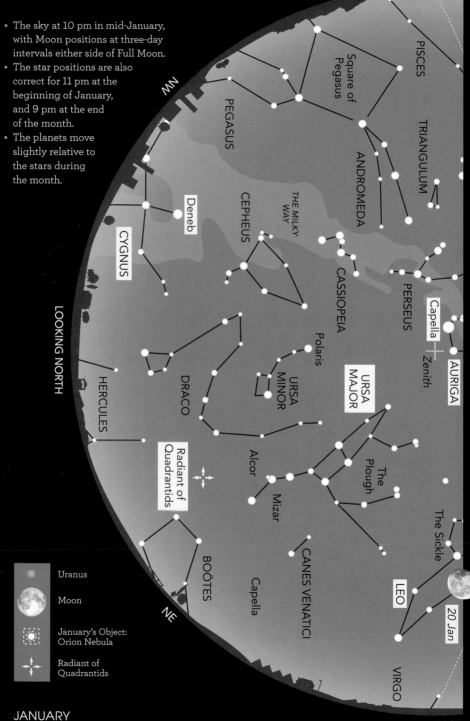

- The sky at 10 pm in mid-January, with Moon positions at three-day intervals either side of Full Moon.
- The star positions are also correct for 11 pm at the beginning of January, and 9 pm at the end of the month.
- The planets move slightly relative to the stars during the month.

WEST

LOOKING NORTH

PISCES

Square of Pegasus

PEGASUS

TRIANGULUM

ANDROMEDA

THE MILKY WAY

CEPHEUS

Deneb

CYGNUS

CASSIOPEIA

PERSEUS

Capella

Zenith

AURIGA

HERCULES

DRACO

Polaris

URSA MINOR

URSA MAJOR

Radiant of Quadrantids

Alcor

The Plough

Mizar

BOÖTES

Capella

CANES VENATICI

The Sickle

LEO

20 Jan

VIRGO

NW

NE

EAST

Uranus

Moon

January's Object:
Orion Nebula

Radiant of
Quadrantids

JANUARY

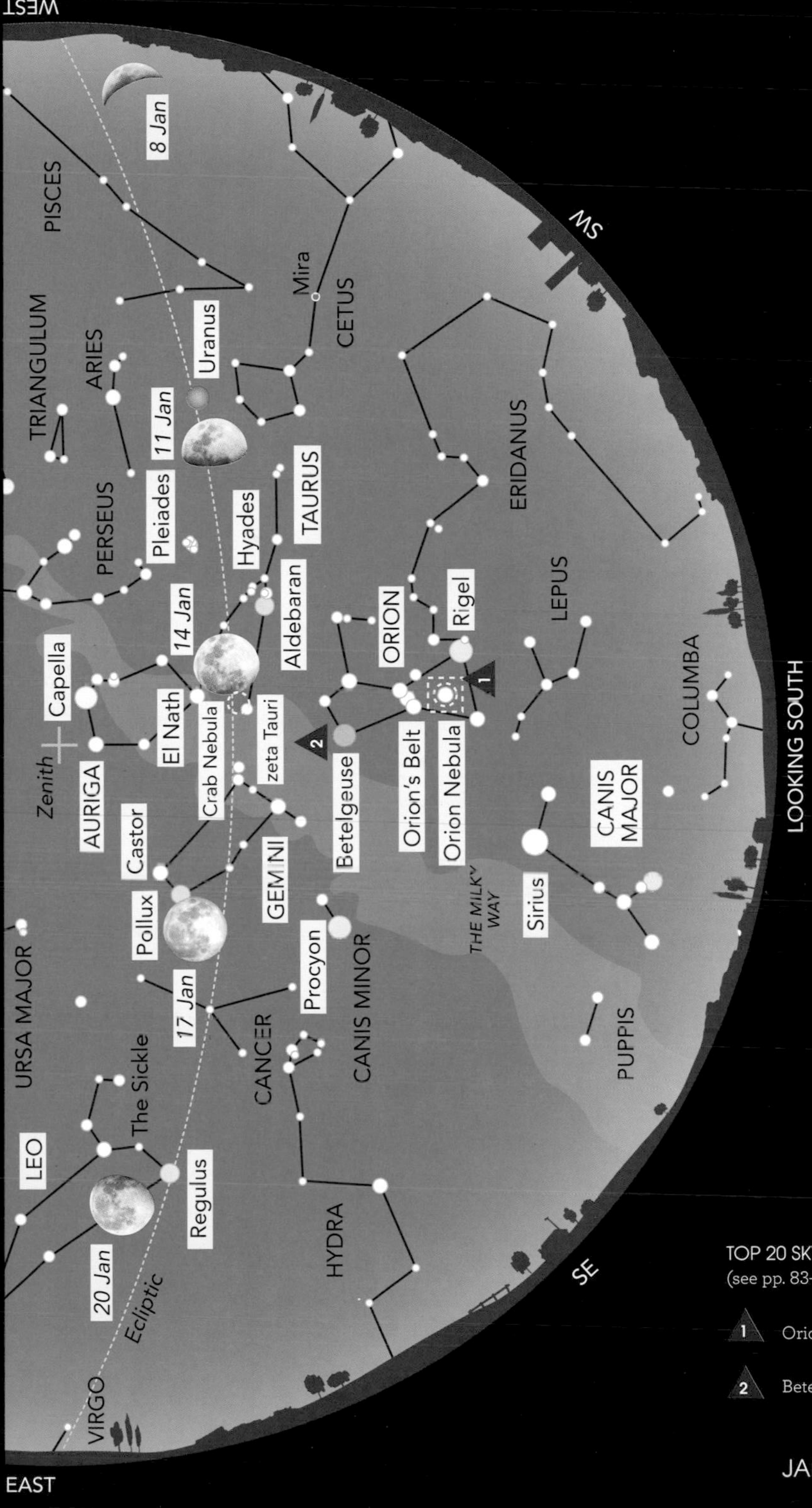

WEST

PISCES

8 Jan

TRIANGULUM

ARIES

Mira

CETUS

MS

PERSEUS

Uranus

11 Jan

Pleiades

Hyades

TAURUS

ERIDANUS

Aldebaran

ORION

Rigel

LEPUS

14 Jan

Capella

El Nath

Crab Nebula

zeta Tauri

Betelgeuse

Orion's Belt

Orion Nebula

COLUMBA

Zenith

AURIGA

Castor

GEMINI

THE MILKY WAY

CANIS MAJOR

LOOKING SOUTH

Pollux

17 Jan

Procyon

Sirius

URSA MAJOR

CANCER

CANIS MINOR

PUPPIS

LEO

The Sickle

Regulus

HYDRA

SE

20 Jan

Ecliptic

VIRGO

TOP 20 SKY SIGHTS
(see pp. 83–85)

1 Orion Nebula

2 Betelgeuse

EAST

The new year opens with a tableau of dazzling stars: **Betelgeuse** and **Rigel** in **Orion**, with glorious **Sirius** in **Canis Major** (the Great Dog) to its lower left. Forming a giant arc above, you'll find **Procyon**, **Castor** and **Pollux** (the celestial twins of **Gemini**), **Capella** and the red giant **Aldebaran**. Also, we are treated to the bright planets Venus, Jupiter, Saturn and Mercury just after sunset, and a brilliant display of shooting stars.

JANUARY'S CONSTELLATION

Taurus (the Bull) is dominated by **Aldebaran**, the baleful eye of the celestial bovine, shining with a magnitude of +0.85. Although its hue is a distinct orange, Aldebaran is classified as a red giant star, a bloated and more elderly version of our Sun.

The 'head' of the bull is formed by the **Hyades** star cluster, 153 light years away. Although Aldebaran looks to be part of this star grouping, it lies at less than half that distance. Even further off (440 light years) is the much more spectacular **Pleiades** (or Seven Sisters) star cluster, packed with very young and brilliant stars (see November's Object).

OBSERVING TIP

Hold a 'meteor party' to check out the year's best celestial firework show, the Quadrantid meteor shower on 3/4 January. You don't need any optical equipment – in fact, telescopes and binoculars will restrict your view of the shooting stars, which can appear anywhere. The ideal viewing equipment is your unaided eye, plus a warm sleeping bag and a lounger. Everyone should look in different directions, so you can cover the whole sky: shout out 'Meteor!' when you see a shooting star. One of the party can record the observations, using a watch, notepad and red torch.

Taurus has two 'horns': the star **El Nath** (Arabic for 'the butting one') to the north, and **zeta Tauri**, whose Babylonian name Shurnarkabti-sha-shutu – meaning 'star in the bull towards the south' – is thankfully not used very much! Nearby, Chinese astronomers witnessed a brilliant supernova in 1054. The remains of this exploding star are visible today through a medium-sized telescope, as the still-glowing **Crab Nebula**.

JANUARY'S OBJECT

Below the distinctive three stars of **Orion's Belt** lies a fuzzy patch, easily visible to the unaided eye under dark skies. Through binoculars or a small telescope you can see that the **Orion Nebula** is a luminous cloud of gas. Some 24 light years across and 1340 light years away, this nebula is part of a vast region of star-birth, the nearest location to Earth where heavyweight stars are being born. The Orion 'maternity ward' contains hundreds of fledgling stars, which have just hatched out of immense dark clouds of dust and gas; there's enough raw material here to make another 100,000 stars.

JANUARY'S TOPIC: STAR NAMES

Just glance at this month's Star Chart, and you'll see that the brightest stars have some pretty exotic names. The

Around 4 am on 21 October 2017, Nigel Bradbury took this spectacular shot using a Canon 6D camera with a Sigma 20-mm lens. The exposure was 3.2 seconds at ISO 1600. See page 87 for more information.

oldest go back over two millennia, when the Greeks named the brightest star **Sirius** after their word for 'the Scorcher' (see February's Object) and **Castor** and **Pollux** from heroic twins in ancient mythology. Latin names generally don't date back to the Romans, but were created by European scholars of the Renaissance. Nicolaus Copernicus, for instance, dubbed **Leo**'s brightest star **Regulus** ('little king').

But the vast majority of star names are Arabic, bestowed on them about a thousand years ago by astronomers in the Middle East who inherited the Greek star catalogues. That's why so many stars begin with the letters 'al' (Arabic for 'the'). Aldebaran, in the constellation Taurus, means 'the follower,' because it tracks behind the Pleiades during the night. **Deneb**, in **Cygnus**, also has Arabic roots, meaning 'the tail (of the flying bird)'.

Oddly named **Betelgeuse** is more of a puzzle. For many years, it was taken to mean 'the armpit of the central one'. But the 'B' in Betelgeuse turned out to be a mistransliteration of an Arabic letter, and the name probably just means 'Orion's hand'.

JANUARY'S PICTURE

Good things come to those who wait, and the old saying certainly held true for Nigel Bradbury when he was in Iceland with an aurora tour group. Lashed by stormy weather, most of the group holed up in their hotels. But Nigel chose to stay outside all night. . . 'My camera was covered in a plastic bag while the rain battered me until 03.00, then a small hole in the clouds appeared, and bright green was seen beyond!'

It was the ethereal glow of the Northern Lights, or aurora, where electrically charged particles from the Sun light up the upper atmosphere. Here, atoms of oxygen are glowing green, the aurora shaped into curtains by the Earth's magnetic field; the clouds are illuminated by the lights of Reykjavik, and in background you can spot the familiar stars of the Plough.

JANUARY'S CALENDAR

SUNDAY	MONDAY	TUESDAY	WEDNESDAY	THURSDAY	FRIDAY	SATURDAY
30	31					1
2 6.33 pm New Moon	3 Quadrantids	4 Quadrantids (am); Earth at perihelion	5 Moon between Jupiter and Saturn	6 Moon near Jupiter	7 Mercury E elongation	8
9 6.11 pm First Quarter Moon	10	11	12 Moon near Ceres and the Pleiades	13 Moon near Ceres and the Pleiades (am); Moon near Aldebaran	14	15
16	17 11.48 pm Full Moon, near Castor and Pollux	18	19 Moon near Regulus	20 Moon near Regulus	21	22
23 Moon near Spica	24 Moon near Spica	25 1.41 pm Last Quarter Moon	26 Moon occults Zubenelgenubi (am)	27	28 Moon near Antares (am)	29 Moon near Venus and Mars (am)

SPECIAL EVENTS

- **Night of 3/4 January:** Maximum of the **Quadrantid meteor shower** – dust particles from the old comet 2003 EH$_1$ burning up in the Earth's atmosphere. It's one of most prolific meteor showers, often featuring coloured shooting stars. With the Moon well out of the way, this year's display should be spectacular.

- **4 January, 6.54 am:** The Earth is at perihelion, its closest point to the Sun (147 million km away).

- **5 January:** You'll find the crescent Moon below Jupiter, with Saturn, Mercury and Venus to the lower right (Chart 1a).

- **6 January:** There's a line of planets to the lower right of the Moon – Jupiter, Saturn, Mercury and Venus.

- **Night of 12/13 January:** The Moon passes half a degree (one Moon-diameter) above dwarf planet Ceres (magnitude +7.1) – catch it with binoculars. The Pleiades lie just above.

- **26 January, 5.20–6.50 am:** The Moon moves in front of the lovely double star Zubenelgenubi (magnitudes +2.7 and +5.2) in Libra (the timing will vary by a few minutes depending on your location in the British Isles).

- **29 January, before dawn:** The crescent Moon lies to the right of Venus, with Mars in between (Chart 1b).

1a 5 January, 5 pm. The Moon, Jupiter, Saturn, Mercury and Venus.

1b 29 January, 6.15 am. Venus, Mars and the crescent Moon.

- On the first few days of the New Year, look very low in the south-west just after sunset to spot the brilliant Evening Star, shining at magnitude −4.3. But **Venus** is slipping rapidly into the twilight, and – after passing between the Earth and Sun on 9 January – it reappears as the Morning Star by mid-January, rising in the south-east about 7 am.

- **Mercury** (magnitude −0.7) lies to the left of Venus at the start of January, and remains low in the evening sky for almost two weeks, fading all the time. At its greatest separation from the Sun on 7 January, Mercury sets about 5.50 pm. The innermost planet is initially rising towards Saturn, and it approaches to almost 3 degrees of the ringworld on 12 January before dropping down into the evening glow.

- **Saturn** lies in Capricornus. Shining at magnitude +0.7, low in the south-west, it's setting about 6 pm and disappears from view in the dusk twilight in the second half of January.

- To the upper left of all these planets, you'll find **Jupiter** in the early evening, resplendent in Aquarius at magnitude −2.1. The giant planet sets around 8 pm.

- **Neptune** (magnitude +7.9) lies on the other side of Aquarius and sinks below the horizon about 9.30 pm. It's followed by **Uranus**, in Pisces, which is a little brighter at magnitude +5.7 and sets around 2 am.

- **Mars** rises about 6 am, in the south-east. It starts the month near Antares, and moves through Ophiuchus and Sagittarius during January. At the close of the month, the Red Planet lies to the lower right of Venus, though it's 250 times fainter at magnitude +1.4.

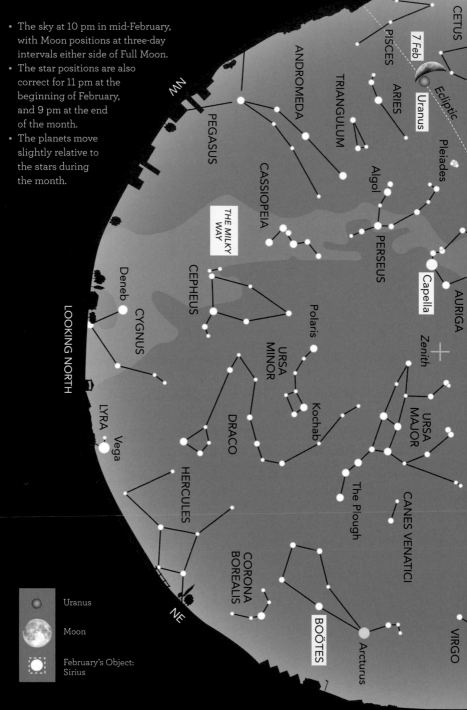

- The sky at 10 pm in mid-February, with Moon positions at three-day intervals either side of Full Moon.
- The star positions are also correct for 11 pm at the beginning of February, and 9 pm at the end of the month.
- The planets move slightly relative to the stars during the month.

WEST

CETUS

PISCES

7 Feb

ARIES

Uranus

Ecliptic

Pleiades

ANDROMEDA

TRIANGULUM

Algol

PERSEUS

Capella

AURIGA

NW

PEGASUS

CASSIOPEIA

THE MILKY WAY

Zenith

Deneb

CEPHEUS

Polaris

URSA MINOR

Kochab

URSA MAJOR

LOOKING NORTH

CYGNUS

DRACO

The Plough

CANES VENATICI

LYRA

Vega

HERCULES

CORONA BOREALIS

BOÖTES

Arcturus

VIRGO

NE

Uranus

Moon

February's Object: Sirius

EAST

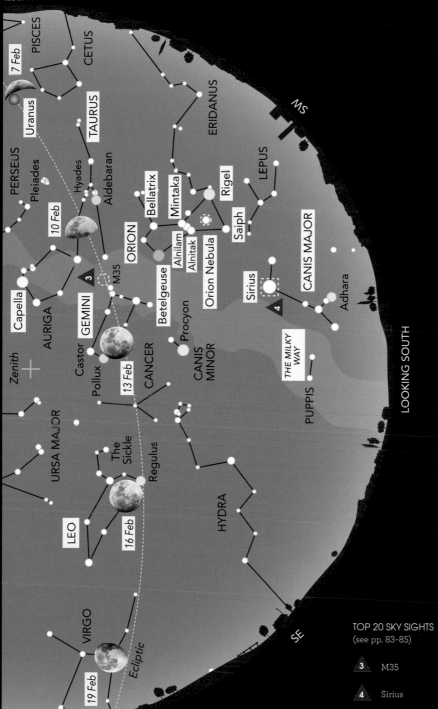

FEBRUARY

WEST

PISCES
CETUS
7 Feb
Uranus
PERSEUS
TAURUS
Pleiades
Hyades
Aldebaran
10 Feb
Capella
AURIGA
Castor
GEMINI
Pollux
13 Feb
CANCER
Zenith
URSA MAJOR
The Sickle
Regulus
LEO
16 Feb
HYDRA
VIRGO
Ecliptic
19 Feb

ORION
Bellatrix
Mintaka
Alnilam
Alnitak
Betelgeuse
Orion Nebula
M35
3
Procyon
CANIS MINOR

ERIDANUS
Rigel
LEPUS
Saiph
Sirius
4
CANIS MAJOR
Adhara
THE MILKY WAY
PUPPIS

MS

SE

LOOKING SOUTH

EAST

TOP 20 SKY SIGHTS
(see pp. 83–85)

3 M35

4 Sirius

FEBRUARY   15

The evenings this month are quiet on the planetary front, but we can still enjoy the bright winter constellations – centred on **Orion, Taurus** and **Gemini** – threaded by the glowing band of the **Milky Way.** These star patterns have drifted to the west since the beginning of January, however, while new constellations are rising higher in the east, led by **Leo** (the Lion) and **Boötes** (the Herdsman). That's because our perspective on the Universe is constantly changing as the Earth orbits the Sun.

FEBRUARY'S CONSTELLATION

Spectacular **Orion** is a rare star grouping that looks like its namesake – a giant hunter with a sword below his belt, wielding a club. The seven main stars lie in the 'Top 70' brightest stars in the sky, but they're not closely associated. Instead, they simply line up, one behind the other.

Closest – at 250 light years – is the fainter of the two stars forming the hunter's shoulders, **Bellatrix.** Next comes the other shoulder-star, blood-red **Betelgeuse**: astronomers are uncertain of its exact distance, but reckon it lies between 550 and 720 light years away. This giant star is a thousand times larger than our Sun, and its fate will be to explode as a supernova.

Around 860 light years from us, slightly brighter blue-white **Rigel** (Orion's foot) is a young star twice as hot as

OBSERVING TIP

Venus is a glorious Morning Star this month, and a small telescope – or even good binoculars – will reveal its striking crescent shape. But you may be disappointed if you observe the planet immediately it rises: seen against a black sky, Venus is so brilliant it's difficult to make out any details. You're best-off waiting until the sky begins to brighten, and the planet appears less dazzling against a pale blue sky.

our Sun, and 125,000 times more luminous. **Saiph,** the hunter's other foot, lies 650 light years away.

We must travel some 1300 light years from home to reach the stars of Orion's glittering belt – **Alnitak, Alnilam** and **Mintaka** – and the great **Orion Nebula.** This huge, glowing cloud of gas is illuminated by brilliant newly born stars. Behind its glitz lies a dark, dusty 'star factory' containing hundreds of embryonic stars, in many cases surrounded by a dusty disc that's poised to form into a system of planets (see January's Object).

FEBRUARY'S OBJECT

Sirius is lording it over the night skies this month, at the head of **Canis Major** (the Great Dog). Though it's our brightest star, at magnitude –1.47, Sirius is not particularly luminous: it just happens to lie nearby, at 8.6 light years.

Popularly known as the Dog Star, it was named Seirios ('the Scorcher') by the ancient Greeks. They believed it added to the Sun's heat in summer, to create the hot, humid 'dog days', when everything – including canines – slowed down. To the Egyptians, the appearance of Sirius at dawn heralded the Nile floods, a welcome harbinger for a bumper harvest.

Boasting a temperature of almost 10,000°C, Sirius is twice as heavy as the

Sun. And it's relatively young: just 230 million years old, compared to the Sun's venerable 4600 million years.

With a 150-mm (or larger) telescope, you can spot a companion star ten thousand times fainter, at magnitude +8.4. Nicknamed 'the Pup', it's a star that has puffed off its outer layers, to end up as a dense white dwarf: an object the same weight as the Sun, and yet only the size of the Earth.

FEBRUARY'S TOPIC: STAR COLOURS

If I asked you what colour the stars are, you'd naturally reply 'white'. But look more carefully at some of the glorious gems of midwinter, with binoculars if you can. And you'll find **Betelgeuse** shines red, while **Capella** is yellow and **Rigel** sparkles in blue-white hues.

These colours are a cosmic thermometer. Amazingly, without knowing even a star's nature or its distance, you can take its temperature. The coolest stars are red, while orange and yellow stars are progressively hotter, with blue-white stars right at the top of the temperature scale.

For instance, Betelgeuse's ruddy hue tells us that its surface temperature is a 'mere' 3300°C, compared to 5500°C for our yellow Sun. This coolish star glows only dully, like an expiring log fire: but its huge size – almost 1000 times wider than the Sun – means its total luminosity is very high. That's why astronomers call it a 'red giant'.

Capella is hotter, so it has a yellow aura, like the Sun. Rigel is near the top end of the stellar temperature scale, its blue-white surface at an incandescent 12,000°C. This super-hot star shines 120,000 times more brilliantly than our Sun.

FEBRUARY'S PICTURE

It's literally a once-in-lifetime sight: Jupiter and its moons visible in the same telescope view as the rings of Saturn. This was the Great Conjunction of 21 December 2020, when Jupiter sailed almost in front of Saturn as seen from the Earth, in their closest encounter for almost 400 years.

While many British astronomers were thwarted by clouds, Damian Peach took advantage of remote observing to image the spectacular event with a telescope in Chile. The clear steady air in the Andes yielded this pin-sharp image of the Solar System's two giants, with Jupiter's four largest moons and its Great Red Spot, along with the ringed planet's biggest moon Titan.

Fortunately, we don't have to wait so long for a repeat of this stunning view: the next Great Conjunction is in 2080.

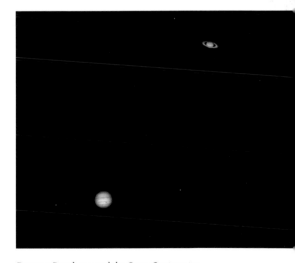

Damian Peach imaged the Great Conjunction with the 1-m f/8 Ritchey–Chrétien reflector at the Chilescope Observatory, using an FLI ProLine 16803 CCD camera and red, green and blue (RGB) filters (more details on page 88).

SUNDAY	MONDAY	TUESDAY	WEDNESDAY	THURSDAY	FRIDAY	SATURDAY
		1 5.46 am New Moon	2 Moon below Jupiter	3 Moon near Jupiter	4	5
6	7 Moon near Uranus	8 1.50 pm First Quarter Moon	9 Moon between the Pleiades and Aldebaran	10	11	12
13 Moon near Castor and Pollux	14	15	16 4.56 pm Full Moon near Regulus; Mercury W elongation	17	18	19
20 Moon near Spica	21	22	23 10.32 pm Last Quarter Moon	24 Moon near Antares (am)	25	26 Moon near Venus and Mars (am)
27	28					

SPECIAL EVENTS

- **2 February:** The thinnest crescent Moon hangs below Jupiter shortly after sunset (Chart 2a).
- **3 February:** Jupiter lies to the lower right of the crescent Moon (Chart 2a).
- **7 February:** It's a good evening to identify faint Uranus, using the fat crescent Moon. With binoculars, follow the terminator on the Moon (the line between the bright and dark regions) to the upper right for three Moon-diameters, to find a speck of light that is the seventh planet.

Uranus

- **26 February, before dawn:** A lovely tableau of the crescent Moon low on the horizon, with Venus and Mars to the left (Chart 2b).

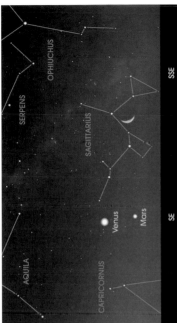

2a 2-3 February, 6.15 pm. The crescent Moon with Jupiter.

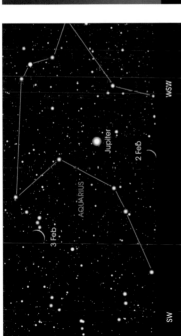

2b 26 February, 6 am. Venus, Mars and the crescent Moon.

- Brilliant **Jupiter** hangs low in the south-west after sunset. Shining at magnitude –2.0 in Aquarius, the giant world sets about 6.30 pm. Around the middle of the month it disappears into the twilight glow.

- Once Jupiter has set, the only planets on view for most of the night are the two faint outermost worlds. **Neptune** (magnitude +7.9) lies in Aquarius and sets about 7.30 pm. **Uranus**, in Aries, sinks below the horizon around midnight, and is on the verge of naked-eye visibility at magnitude +5.8.

- We need to turn to the pre-dawn sky for more planetary action. Glorious **Venus** rises in the south-east about 5 am; at magnitude –4.6 it far outshines every star.

- **Mars**, in Sagittarius, starts the month to the lower right

of Venus but considerably fainter at magnitude +1.3. The Red Planet clears the horizon around 5.30 am. Mars is moving to the left across the sky, and lies below Venus during the second half of February.

- **Mercury** is deep in the dawn twilight to the lower left of Venus, its brightness increasing throughout the month from an inconspicuous magnitude

+1.2 to –0.1. The innermost planet rises about 6.15 am and reaches its greatest separation from the Sun on 16 February.

- At the end of the month you may just spot **Saturn** rising at 6.15 am, to the left of Mercury. The ringed planet lies in Capricornus and it's a bit fainter than Mercury, at magnitude +0.8.

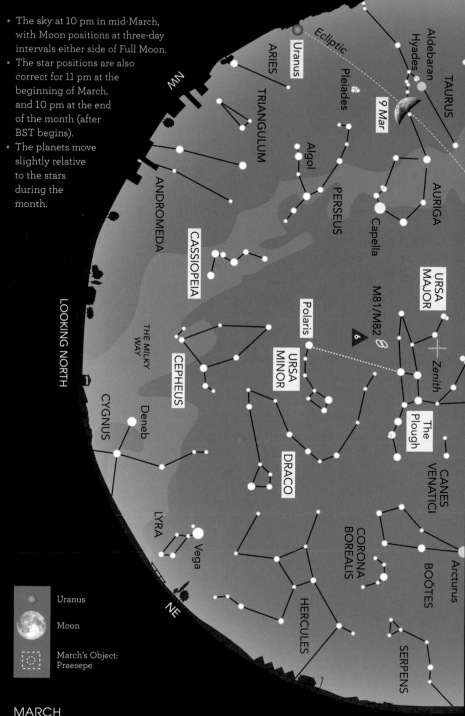

- The sky at 10 pm in mid-March, with Moon positions at three-day intervals either side of Full Moon.
- The star positions are also correct for 11 pm at the beginning of March, and 10 pm at the end of the month (after BST begins).
- The planets move slightly relative to the stars during the month.

WEST

Ecliptic

Uranus

ARIES

Aldebaran

Hyades

TAURUS

Pleiades

9 Mar

NW

TRIANGULUM

Algol

AURIGA

PERSEUS

Capella

ANDROMEDA

CASSIOPEIA

URSA MAJOR

M81/M82

Polaris

LOOKING NORTH

THE MILKY WAY

CEPHEUS

URSA MINOR

Zenith

CYGNUS

Deneb

DRACO

The Plough

CANES VENATICI

LYRA

Vega

CORONA BOREALIS

BOÖTES

Arcturus

NE

HERCULES

SERPENS

Uranus

Moon

March's Object: Praesepe

EAST

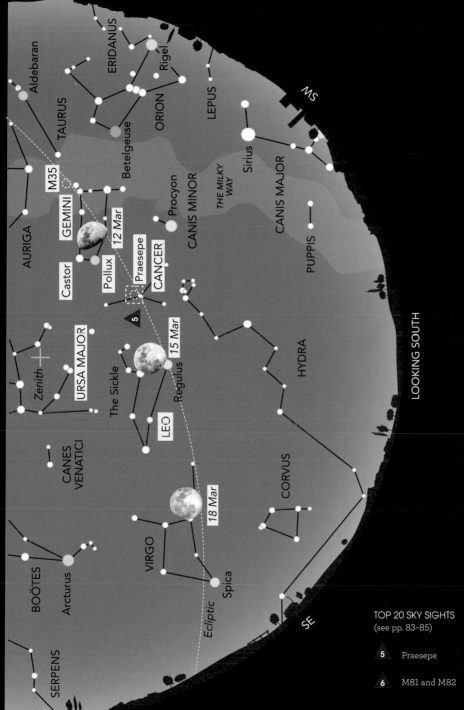

MARCH

WEST

TAURUS
Aldebaran
ERIDANUS
Rigel
ORION
Betelgeuse
LEPUS
MS
Sirius
CANIS MAJOR
PUPPIS

AURIGA
M35
GEMINI
Castor
Pollux
12 Mar
Procyon
CANIS MINOR
THE MILKY WAY
Praesepe
CANCER
5

URSA MAJOR
Zenith
The Sickle
15 Mar
Regulus
LEO
HYDRA

CANES VENATICI

CORVUS

18 Mar

VIRGO

BOÖTES
Arcturus
Spica
Ecliptic

SERPENS

SE

LOOKING SOUTH

TOP 20 SKY SIGHTS
(see pp. 83–85)

5   Praesepe

6   M81 and M82

Winter is officially over on 20 March, as we pass the equinox and day becomes longer than the night. Though most of the changing action in the sky happens towards the south, we shouldn't forget the northern constellations that are visible every night of the year: Queen **Cassopeia** and her consort **Cepheus**, **Draco** (the Dragon), and the two Bears – **Ursa Major** and **Ursa Minor**.

MARCH'S CONSTELLATION

The constellation of the celestial Twins – **Gemini** – is crowned by the bright stars **Castor** and **Pollux**, representing the heads of the celestial brethren, with their bodies running in parallel lines of stars. In legend they were conceived by princess Leda on the same day, Castor by her human husband, and immortal Pollux by chief god Zeus. To prevent the devoted pair from being separated by Castor's death, Zeus placed them together for eternity among the stars.

Castor is an amazing family of six stars. Through a small telescope, you can see that it's a close double, with a fainter companion lying further away. And all three of these stars are themselves very tight-knit pairs. Pollux is a cooler orange giant. Though it is a single star, Pollux is not entirely alone: it's orbited by a planet – Thestias – that's mightier than Jupiter.

Gemini also boasts a pretty star cluster, **M35**. Even at a distance of nearly 2800 light years, it's visible to the unaided eye, and a fine sight through binoculars or a small telescope.

MARCH'S OBJECT

'A nebulous mass in the breast of the Crab' is how the ancient astronomer Ptolemy described this month's object. In 1609, Galileo sought it out with his new

OBSERVING TIP

This is the ideal time of year to tie down the main compass directions, as seen from your observing site. North is easy – just latch onto Polaris, the Pole Star, using the familiar stars of the Plough (see June's Constellation). And at noon, the Sun is always in the south. But the useful extra in March is that we hit the Spring Equinox, when the Sun rises due east, and sets due west. So remember those positions relative to a tree or house around your horizon.

telescope, revealing it consists of at least 40 stars. We now know that **Praesepe** (literally 'the manger') is a swarm of over 1000 stars, prompting the more poetic nickname the Beehive Cluster.

Praesepe lies 600 light years away, and its stars were born together some 600 million years ago. Two of its stars have planets in orbit about them, but they are not 'Earths' – instead, they are 'hot Jupiters', gas giants circling close in to their parent star.

With the naked eye, you can spot Praesepe as a faint misty patch in **Cancer**, between **Gemini** and **Leo**. It is a magnificent sight through binoculars or a small telescope, its stars scattered like jewels on black velvet and spread out over an area three times wider than the Moon.

MARCH'S TOPIC: JOHANNES KEPLER

NASA's space telescope Kepler was most aptly named. The 17th-century astronomer Johannes Kepler (1571–1630) worked out how our local planets orbit the Sun, while his space-faring namesake was designed to discover planets orbiting other stars.

Born in Germany, Kepler was the leading mathematician of his day. At that time, most scientists agreed that the planets move around the Sun, rather than around the Earth. But they were fixated by the idea that the planets' paths were circles – and that just didn't fit the observations. Kepler spent years struggling with this problem, until one day he proposed squeezing the planets' orbits 'like a German sausage'. Eureka! The planets did indeed move in oval paths, called ellipses.

Kepler's Laws, concerning the shape of the planets' orbits and how fast they move, are still fundamental to astronomy and space exploration today.

But Kepler accomplished much more. He devised a telescope that was better than his contemporary Galileo (1564–1642) was using, and he observed a supernova that exploded in 1604. Kepler also wrote the first science-fiction novel, about humans visiting the Moon. And he did all this while defending his mother from charges of witchcraft – and securing her acquittal.

MARCH'S PICTURE

As the Earth spins around in space, most of the stars seem to move across the sky in the opposite direction. Some rise in the east and set in the west; others wheel around the heavens. But one star doesn't partake in this celestial choreography. **Polaris**, the Pole Star, is practically stationary because the Earth's axis is pointing towards it.

Jo Bourne's long-exposure image shows clearly how stars trace out long arcs around the celestial pole over the course of an hour, above a lone tree at Cissbury Ring in West Sussex. The exception is almost unmoving Polaris, that appears as a central spot of light.

Using a Canon 60D camera at f/2.8 and ISO 400, Jo Bourne took 114 × 30-second exposures to create this image. See page 87 for more information.

SUNDAY	MONDAY	TUESDAY	WEDNESDAY	THURSDAY	FRIDAY	SATURDAY
		1	2 5.35 pm New Moon	3	4	5
6	7	8 Moon between the Pleiades and Aldebaran	9	10 10.45 am First Quarter Moon	11	12 Moon near Castor and Pollux
13 Moon near Castor and Pollux	14	15 Moon near Regulus	16	17	18 7.17 am Full Moon	19 Moon near Spica
20 Spring Equinox; Venus W elongation	21	22	23 Moon near Antares (am)	24 Moon near Antares (am)	25 5.37 am Last Quarter Moon	26
27 BST begins	28 Moon below Venus, Saturn anc Mars	29	30	31		

SPECIAL EVENTS

• **8 March:** The Moon lies near the Pleiades, with Aldebaran to the left (Chart 3a).

• **20 March, 3.33 pm:** The Spring Equinox, when day and night are equal.

• **27 March, 1.00 am:** British Summer Time starts – don't forget to put your clocks forward.

• **28 March, before dawn:** Low in the south-east before sunrise, you may catch Venus with the narrow crescent Moon below. Use binoculars to pick out Saturn and Mars in between them (Chart 3b).

Venus

3a *8 March, 10 pm. The Moon between Aldebaran and the Pleiades.*

3b *28 March, 6 am. The crescent Moon with Venus, Saturn and Mars.*

• There are no bright planets in the sky almost all night long, only **Uranus**, which – at magnitude +5.8 – is hardly visible to the unaided eye. The seventh planet lies in Aries, and sets around 10.30 pm.

• To spot the other worlds of our Solar System, you'll have to stay awake till 4.30 am, when **Venus** rises in the south-east. The Morning Star blazes at a brilliant magnitude –4.4, and reaches its maximum separation from the Sun on 20 March. Through a small telescope you'll see it as a thick crescent which broadens as the month progresses and becomes exactly half illuminated (technically known as dichotomy) on 21 March.

• **Mars** is below and to the right of Venus, and over a hundred times fainter at magnitude +1.1. Lying in Capricornus, the Red Planet rises around 5 am.

• For the first few nights of March, **Mercury** (magnitude –0.1) lies well to the left of this pair, low in the dawn twilight and rising just after 6 am. Just above it is Saturn, at magnitude +0.8. As the month progresses, Mercury sinks from view and Saturn moves upward in the sky.

• Mid-month, **Saturn** – in Capricornus – is rising at 5.30 am. By the last week of March, the ringworld is lying near Venus and Mars, with the two fainter planets completing a triangle below the Morning Star (Chart 3b).

• **Jupiter** and **Neptune** are lost in the Sun's glare this month.

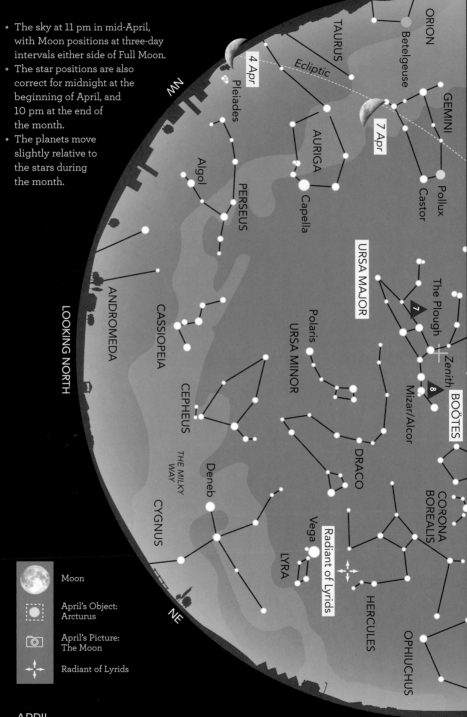

- The sky at 11 pm in mid-April, with Moon positions at three-day intervals either side of Full Moon.
- The star positions are also correct for midnight at the beginning of April, and 10 pm at the end of the month.
- The planets move slightly relative to the stars during the month.

WEST

ORION
Betelgeuse
TAURUS
4 Apr
Ecliptic
Pleiades
GEMINI
7 Apr
Algol
AURIGA
Pollux
Castor
NW
PERSEUS
Capella
URSA MAJOR
ANDROMEDA
The Plough
7
CASSIOPEIA
Polaris
Zenith
Mizar/Alcor
8
BOÖTES
LOOKING NORTH
URSA MINOR
CEPHEUS
DRACO
CORONA BOREALIS
THE MILKY WAY
Deneb
CYGNUS
Vega
Radiant of Lyrids
LYRA
HERCULES
OPHIUCHUS
NE

Moon

April's Object: Arcturus

April's Picture: The Moon

Radiant of Lyrids

EAST

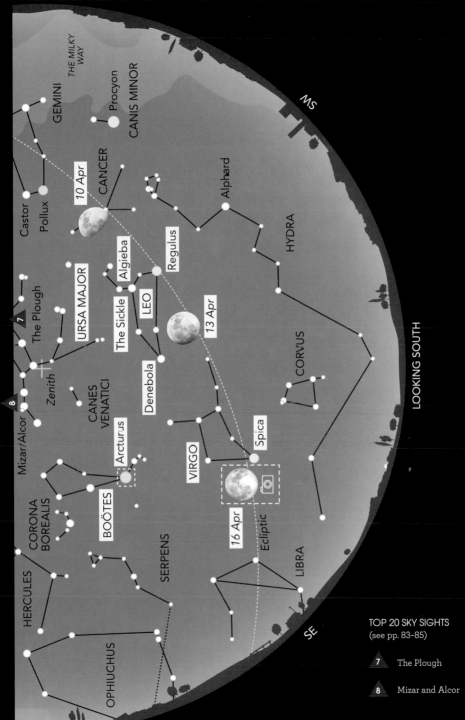

WEST

THE MILKY WAY

GEMINI

Procyon
CANIS MINOR

SW

10 Apr
CANCER

Castor
Pollux

Algieba

Regulus

Alphard

HYDRA

URSA MAJOR

The Sickle

LEO

13 Apr

7 The Plough

Zenith

Denebola

CANES
VENATICI

CORVUS

Mizar/Alcor

8

Arcturus

Spica

LOOKING SOUTH

CORONA
BOREALIS

BOÖTES

VIRGO

16 Apr

Ecliptic

HERCULES

SERPENS

LIBRA

OPHIUCHUS

SE

EAST

TOP 20 SKY SIGHTS
(see pp. 83–85)

7 The Plough

8 Mizar and Alcor

Three bright stars ride high this month, dominating the major constellations of the spring skies. Leading the way is **Regulus** in **Leo**, with **Virgo's** leading star **Spica** to the lower left, and orange **Arcturus** in **Boötes** lying above. We are treated to a meteor display on 21 April, some interesting planetary action in the dawn skies and Mercury's best evening display at the end of the month.

APRIL'S CONSTELLATION

Like the fabled hunter Orion, **Leo** is one of the rare constellations that resembles the real thing – in this case, a crouching lion. Leo commemorates a fearsome beast that Hercules slaughtered as the first of his 12 labours. The lion's fur couldn't be pierced by any weapon, so Hercules wrestled with the creature and choked it to death.

The lion's heart is marked by the first-magnitude star **Regulus**. This celestial whirling dervish spins around in just 16 hours, making its equator bulge remarkably. Rising upwards is 'the Sickle', a back-to-front question mark that delineates the front quarters, neck and head of Leo. A small telescope shows that **Algieba**, the star that makes up the lion's shoulder, is a beautiful close double star.

The other extremity of Leo is marked by **Denebola**, meaning 'the lion's tail' in Arabic. Just underneath the feline's tummy is a clutch of spiral galaxies; too faint to be seen with the unaided eye, you can track them down with a small telescope.

APRIL'S OBJECT

When orange **Arcturus** appears in the evening sky, it's a sure sign that summer is on the way. The fourth-brightest star in the heavens, Arcturus was a navigational beacon for Polynesian sailors, because it passes directly over Hawaii. Its Greek name means 'the bear guardian', because Arcturus always follows **Ursa Major** (the Great Bear) around the sky.

The brightest star in **Boötes** (the Herdsman), Arcturus is just entering its red giant phase, as it runs out of nuclear fuel at its core. Not much heavier than the Sun, Arcturus has grown to 25 times our star's width. Outshining the Sun almost 200 times over, Arcturus fluctuates slightly in brightness as the star becomes unstable towards the end of its life.

APRIL'S TOPIC: REDSHIFT

In 1914, American astronomer Vesto Slipher (1875–1969) received a standing ovation at the American Astronomical Society for a puzzling new discovery. He had measured the speed of 15 spiral galaxies, and found that most of them are moving away, indicating 'a general fleeing from us or the Milky Way'.

OBSERVING TIP

Don't think that you need a telescope to bring the heavens closer. Binoculars are excellent – and you can fling them into the back of the car at the last minute. For astronomy, buy binoculars with large lenses coupled with a modest magnification. An ideal size is 7 × 50, meaning that the magnification is seven times, and that the diameter of the lenses is 50 millimetres.

Slipher had broken down the light from faint galaxies into a spectrum of colours, or wavelengths, and found that dark bands in the spectrum (absorption lines) were shifted away from their normal position. In most galaxies it was towards the red end of the spectrum: hence the term 'redshift'. He had concluded that a galaxy's redshift was caused by its motion away from us, stretching out its light waves to longer (redder) wavelengths, just as the pitch of an ambulance siren drops as it rushes past us and its sound waves are stretched to a lower pitch.

Fellow American Edwin Hubble (1889–1953) later found that a galaxy's speed depends on its distance. And, in 1927, Belgian astronomer Georges Lemaître (1894–1966) realised this means that galaxies are all rushing apart from a gigantic explosion in which the Universe began – the instant we now call the Big Bang.

APRIL'S PICTURE

What colour is the Moon? It looks pure white, illuminating our night skies with its soft silvery glow. But appearances are deceptive. The Moon is actually grey, reflecting just 12 per cent of the sunlight falling on it – as dark as the tarmac on a car park. It only appears bright in contrast to the blackness of space.

But there are shades of colour in the Moon's grey landscape, clearly visible in this image where James Harrop has enhanced our satellite's natural hues. The brightest regions, the lunar highlands, are replete with a calcium-rich rock called anorthosite. The darker low-lying plains (known as 'maria', meaning 'seas') have filled with lava that's enriched with either iron (orange-brown) or titanium (blue).

The Full Moon on 1 October 2020 (the Harvest Moon) was captured by James Harrop using a Canon 6D camera and 600-mm lens, with an exposure of 1/800 second at ISO 200.

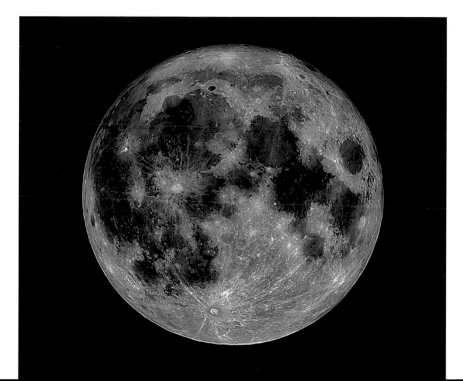

SUNDAY	MONDAY	TUESDAY	WEDNESDAY	THURSDAY	FRIDAY	SATURDAY
					1 7.24 am New Moon	2
3 Moon near Saturn (am)	4 Moon near the Pleiades	5 Mars near Saturn (am); Moon between Pleiades and Aldebaran	6	7	8	9 7.48 am First Quarter Moon near Castor and Pollux
10	11 Moon near Regulus	12 Moon near Regulus	13 Jupiter near Neptune (am)	14	15	16 7.55 pm Full Moon near Spica
17	18	19	20 Moon near Antares (am)	21 Lyrids	22 Lyrids (am)	23 12.56 pm Last Quarter Moon
24 Moon near Saturn (am)	25 Moon between Saturn and Mars (am)	26 Moon between Mars and Venus (am)	27 Moon near Venus and Jupiter (am)	28 Venus near Neptune (am)	29 Mercury E elongation, near the Pleiades	30 Venus near Jupiter (am); 9.28 pm New Moon; partial solar eclipse; Mercury near the Pleiades

SPECIAL EVENTS

- **4–5 April:** A lovely tableau in the evening sky as the crescent Moon sails past the Pleiades and Aldebaran.
- **5 April, before dawn:** Mars passes below Saturn (see Planet Watch and Chart 4a), to the right of brilliant Venus.
- **Night of 21/22 April:** It's an excellent year for observing the maximum of the **Lyrid meteor shower**, as the Moon doesn't rise until 3.30 am. These shooting stars appear to emanate from the constellation Lyra as debris from Comet Thatcher burns up in the Earth's atmosphere, and they often leave glowing trails of dust.
- **24–27 April, before dawn:** The crescent Moon passes below the planets Saturn, Mars, Venus and Jupiter. You'll need a very clear horizon to the east, and binoculars will reveal the planets more clearly in the twilight glow.
- **30 April:** Mercury passes just to the left of the Pleiades (see Planet Watch and Chart 4b).
- **30 April:** A partial eclipse of the Sun is visible from south-western South America and adjacent regions of the Pacific Ocean. Nothing is visible from the British Isles.

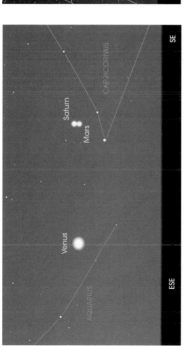

4a 5 April, 5.45 am. Saturn and Mars very close, with Venus to the left.

4b 30 April, 9.30 pm. Mercury lies just to the left of the Pleiades.

- Faint **Uranus** is the only planet keeping watch over the evening sky at the start of April. At magnitude +5.9, it's hanging about in Aries and sets around 9.30 pm.

- Uranus is joined later in the month by **Mercury**, putting on its best evening appearance of the year. The innermost planet appears above the western horizon just after sunset around 10 April, at magnitude −1.5. Fading all the time, Mercury shoots upwards to pass Uranus on 17 April. At its greatest separation from the Sun on 29 April, Mercury has dimmed to magnitude +0.4 and is setting as late as 10.30 pm. On the last two nights of April, Mercury is a lovely sight right next to the Pleiades.

- **Venus** is the undisputed queen of the morning sky, at a resplendent magnitude −4.2 and rising around 5 am. At the star of April, Saturn (magnitude +0.9) and Mars (magnitude +1.1) lie just to the right of Venus, in Capricornus.

- As the days go by, Venus and Mars both move to the left in the sky, with the Red Planet passing just 20 arcminutes under Saturn on the morning of 5 April (Chart 4a). By mid-month, Saturn is rising at 4.20 am and Mars at 4.40 am.

- From about 7 April, you'll be able to spot **Jupiter** rising in the east around 5.30 am, well to the lower left of Venus. Lying in Pisces, the giant planet is second only to the Morning Star in brightness, at magnitude −2.1. The two brilliant worlds are converging, with Jupiter lying just to the left of Venus before dawn on 30 April.

- **Neptune** also lies in Pisces, rising about 5.30 am. The outermost planet is close to Jupiter on 13 April and Venus on 28 April but it will be difficult to spot this faint world (magnitude +7.9) against the bright twilight glow.

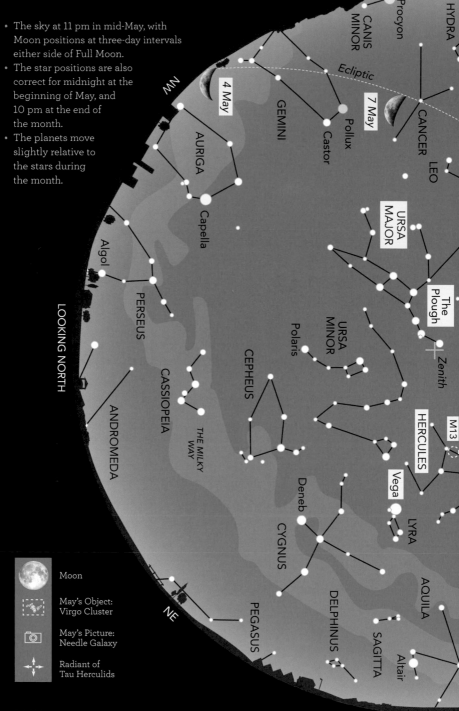

WEST

- The sky at 11 pm in mid-May, with Moon positions at three-day intervals either side of Full Moon.
- The star positions are also correct for midnight at the beginning of May, and 10 pm at the end of the month.
- The planets move slightly relative to the stars during the month.

HYDRA

Procyon

CANIS MINOR

Ecliptic

4 May

7 May

GEMINI

Pollux

CANCER

Castor

LEO

AURIGA

URSA MAJOR

Capella

Algol

The Plough

Zenith

PERSEUS

URSA MINOR

HERCULES

M13

LOOKING NORTH

Polaris

CASSIOPEIA

CEPHEUS

Vega

ANDROMEDA

THE MILKY WAY

LYRA

Deneb

CYGNUS

AQUILA

Moon

DELPHINUS

May's Object:
Virgo Cluster

PEGASUS

SAGITTA

May's Picture:
Needle Galaxy

Altair

Radiant of
Tau Herculids

NE

EAST

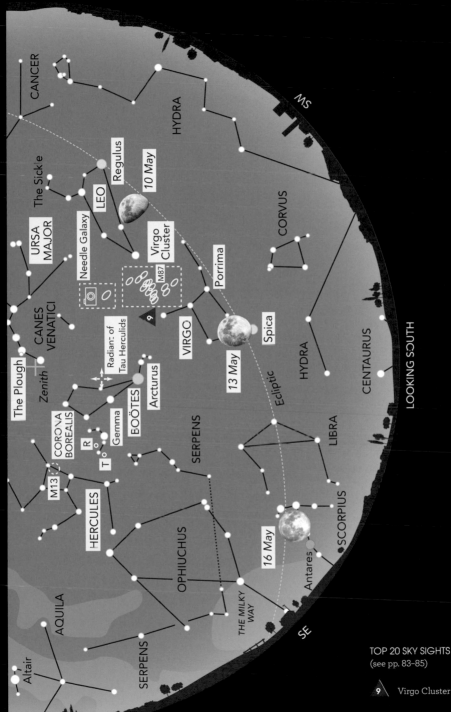

SW

MAY

CANCER

HYDRA

The Sickle

Regulus

10 May

LEO

URSA MAJOR

Needle Galaxy

Virgo Cluster

M87

CORVUS

Porrima

CANES VENATICI

Radiant of Tau Herculids

9

VIRGO

Spica

13 May

HYDRA

The Plough

Zenith

Arcturus

BOÖTES

Ecliptic

CORONA BOREALIS

Gemma

SERPENS

CENTAURUS

R

T

LIBRA

M13

HERCULES

OPHIUCHUS

16 May

SCORPIUS

AQUILA

SERPENS

THE MILKY WAY

Antares

SE

LOOKING SOUTH

Altair

TOP 20 SKY SIGHTS
(see pp. 83–85)

9 Virgo Cluster

This month is packed full of celestial action, though it will help if you're an early bird. The highlights are close conjunctions of Jupiter with Venus and Mars, and a total lunar eclipse; but we are also treated to Mercury with the Pleiades, meteors from Halley's Comet – and possibly a storm of shooting stars at the end of May.

MAY'S CONSTELLATION

The small but perfectly formed Northern Crown (**Corona Borealis**) lies near brilliant **Arcturus**. In legend, it was a wedding gift from Bacchus, the god of wine, to princess Ariadne. The jewel in the crown is blue-white **Gemma** (magnitude +2.2). Along with most of the members of the **Plough**, Gemma is part of the **Ursa Major** Moving Group of stars travelling together through space.

The constellation boasts two remarkable variable stars. Normally at the limit of naked-eye visibility (magnitude +6), **R** Coronae Borealis can unpredictably drop to magnitude +14 when sooty clouds form in its atmosphere and obscure the star's light. **T** Coronae Borealis (the Blaze Star) behaves in the opposite way. It usually skulks at magnitude +11, and then suddenly flares to magnitude +2, as gas from the star falls onto a companion white dwarf and explodes. This 'recurrent nova' last erupted in 1946.

MAY'S OBJECT

If you have a small telescope, sweep the 'bowl' formed by **Virgo's** 'Y' shape, and you'll detect dozens of fuzzy blobs. These are just the brightest members of the **Virgo Cluster**, our closest giant galaxy cluster, some 54 million light years away.

Large telescopes reveal 2000 galaxies in this vast swarm of star cities. Many of them are spirals like our Milky Way, while others are massive elliptical galaxies. The king of the cluster is **M87**, a giant elliptical with a massive central black hole that is ejecting a 5000-light-year-long jet of high-speed subatomic particles.

The Virgo Cluster is so massive that its gravity holds many neighbouring galaxy clusters in thrall – including the Local Group that contains the Milky Way – making it the centre of the Virgo Supercluster that's over 110 million light years across.

MAY'S TOPIC: GLOBULAR CLUSTERS

On a really dark night, look carefully between the bright stars **Vega** and Arcturus, and slightly closer to Vega you'll see a fuzzy patch (binoculars will help).

OBSERVING TIP

It's always fun to search out 'faint fuzzies' in the sky – star clusters, nebulae and galaxies. But don't even think of observing these objects around the time of Full Moon (especially the supermoon this month), as its light will drown them out. You'll have the best views near New Moon, a period astronomers call 'dark of Moon'. When the Moon is bright, though, there's still plenty to see: focus on planets, bright double stars – and, of course, the Moon itself. Check the month-by-month Calendar for the Moon's phases.

This is the Great Cluster in **Hercules**, more often known by its catalogue number **M13**, and it's a closely knit ball of around a million stars.

M13 is just one of over 150 globular clusters that swarm around our Milky Way. Globular clusters surround all major galaxies: the more massive the galaxy, the bigger its entourage. About 12,000 globular clusters orbit the giant galaxy M87 in Virgo.

Globular clusters are very old, consisting of stars born soon after the Big Bang. And they are very densely packed. The average separation between their residents is just one light year (for comparison, we are 4.3 light years from the Sun's nearest neighbour, Proxima Centauri). If we lived in M13, the night sky would be jam-packed with nearby bright stars.

MAY'S PICTURE

Known as the **Needle Galaxy** – for obvious reasons! – this sliver of light in Coma Berenices (a constellation too faint to appear on the Star Chart) is easily visible

Sara Wager used an Orion Optics ODK10 250-mm Dall-Kirkham reflector, with a QSI 683 CCD camera and Baader LRGB filters. She took 50 × 800-second exposures for luminance, and 24 × 600-second exposures through red, green and blue filters. The total exposure time was 37 hours 30 minutes. See pages 88–89 for more information.

in a small telescope as a glowing shard. It's magnificent in this image taken by Sara Wager, a British astronomer observing under the clear skies of Spain.

At magnitude +10, the Needle Galaxy is brighter than many of the 'M' objects in Charles Messier's eighteenth-century list of fuzzy sky sights. Sometimes called 'the celestial masterpiece that Messier missed', this lovely galaxy languishes under the catalogue number NGC 4565.

Lying 50 million light years away, the Needle is a spiral galaxy that's a near twin to our own Milky Way. By chance, we happen to observe NGC 4565 almost edge-on, and you can see the extreme thinness of the disc of a spiral galaxy like ours.

SUNDAY	MONDAY	TUESDAY	WEDNESDAY	THURSDAY	FRIDAY	SATURDAY
1 Venus very near Jupiter (am); Mercury near the Pleiades	**2** Moon and Mercury near the Pleiades	**3**	**4**	**5**	**6** Eta Aquarids (am); Moon near Castor and Pollux	**7**
8	**9** 1.21 am First Quarter Moon near Regulus	**10**	**11**	**12**	**13** Moon occults Porrima (am); Moon near Spica	**14**
15	**16** 5.14 am Full Moon, supermoon; total lunar eclipse	**17** Moon near Antares (am)	**18** Mars near Neptune (am)	**19**	**20**	**21**
22 7.43 pm Last Quarter Moon, near Saturn (am)	**23**	**24**	**25** Moon near Jupiter and Mars (am)	**26** Moon between Jupiter and Venus (am)	**27** Moon very close to Venus (am)	**28**
29 Jupiter near Mars (am)	**30** Jupiter near Mars (am), 12.30 pm New Moon	**31** Tau Herculids				

SPECIAL EVENTS

- **1 May, before dawn:** The two brightest planets, Venus and Jupiter, are very close together (see Planet Watch and Chart 1a).
- **2 May:** Very low in the north-west, about 10 pm, binoculars will reveal the narrow crescent Moon with Mercury and the Pleiades (see Planet Watch).
- **Morning of 6 May:** Shooting stars from the Eta Aquarid meteor shower – tiny pieces of Halley's Comet burning up in Earth's atmosphere – fly across the sky in the early hours of the morning.
- **13 May, 1.55–2.45 am:** The Moon moves in front of **Porrima** (magnitude +2.7) in Virgo (the timing varies by a few minutes depending on your location).

- **16 May:** A total lunar eclipse is visible from the Americas and parts of Africa and Europe. From the British Isles, the partial phase starts at 3.27 am and totality at 4.29 am.
- **25–27 May, before dawn:** Very low in the east, the crescent Moon glides below Jupiter and Mars (25 May) and Venus (27 May): best seen in binoculars.

- **29 May, before dawn:** Mars passes just below Jupiter (see Planet Watch and Chart 5b).
- **Night of 31 May/ 1 June:** Debris from Comet Schwassman-Wachman-3 may produce a storm of shooting stars, the **Tau Herculid meteor shower** (the radiant has now moved to Boötes). Expect the best views before dawn.

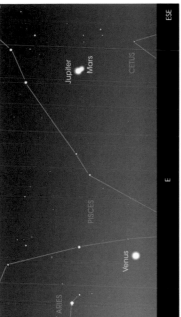

5a 1 May, 5 am. Jupiter and Venus very close, with Mars and Saturn.

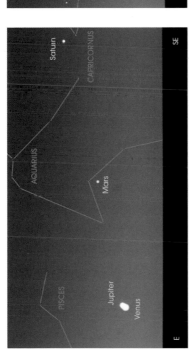

5b 29 May, 4 am. Jupiter and Mars very close, with Venus.

• On the evening of 1 May, **Mercury** lies just to the left of the Pleiades, and the following night they are joined by a narrow crescent Moon. The innermost planet shines at magnitude +0.7, just a tad brighter than Aldebaran to the left. Setting about 10 pm, Mercury fades rapidly as the month progresses, and disappears into the twilight glow by the middle of the May.

• The next planet that's visible is **Saturn**, rising in the south-east around 2.30 am. At magnitude +0.8, the ringed planet is to be found in Capricornus.

• Just before dawn, an intriguing planetary dance takes place during May. The brightest of these worlds is **Venus**, at magnitude −4.0. Rising about 4 am, the Morning Star is tracking downwards and to the left. On the morning of 1 May, we're treated to the stunning sight of the Morning Star up close and personal to the second-most brilliant planet, as Venus passes below **Jupiter** at a distance of just 20 arcminutes (Chart 5a).

At magnitude −2.1, the giant planet resides in Pisces all month, and rises around 3.30 am in the middle of May.

• **Neptune** lies to the right of Jupiter, smack on the border of Pisces and Aquarius. Also rising about 3.30 am, the most distant planet shines a feeble magnitude +7.9.

• At the start of May, you'll find **Mars** well to the right of these worlds, in the centre of Aquarius (Chart 5a). As the month progresses, the Red Planet (magnitude +0.8) tracks rapidly to the left and moves into Pisces, passing Neptune on 18 May. It's just 40 arcminutes from much brighter Jupiter on 29 May, both planets rising around 3.30 am (Chart 5b).

• **Uranus** is lost in the Sun's glare this month.

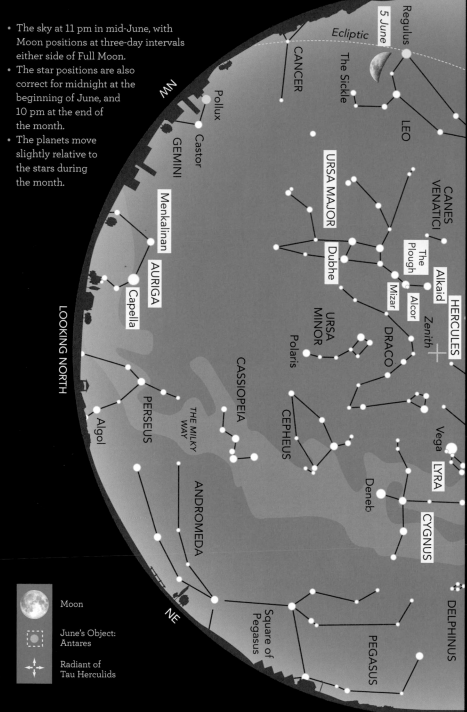

- The sky at 11 pm in mid-June, with Moon positions at three-day intervals either side of Full Moon.
- The star positions are also correct for midnight at the beginning of June, and 10 pm at the end of the month.
- The planets move slightly relative to the stars during the month.

WEST

Regulus

5 June

Ecliptic

CANCER

The Sickle

LEO

NW

Pollux

Castor

GEMINI

CANES VENATICI

Menkalinan

URSA MAJOR

AURIGA

Dubhe

The Plough

Alkaid

HERCULES

Capella

Mizar

Alcor

Zenith

Mizar

LOOKING NORTH

URSA MINOR

DRACO

Polaris

CASSIOPEIA

Vega

PERSEUS

THE MILKY WAY

CEPHEUS

LYRA

Algol

Deneb

CYGNUS

ANDROMEDA

Square of Pegasus

NE

DELPHINUS

PEGASUS

Moon

June's Object:
Antares

Radiant of
Tau Herculids

EAST

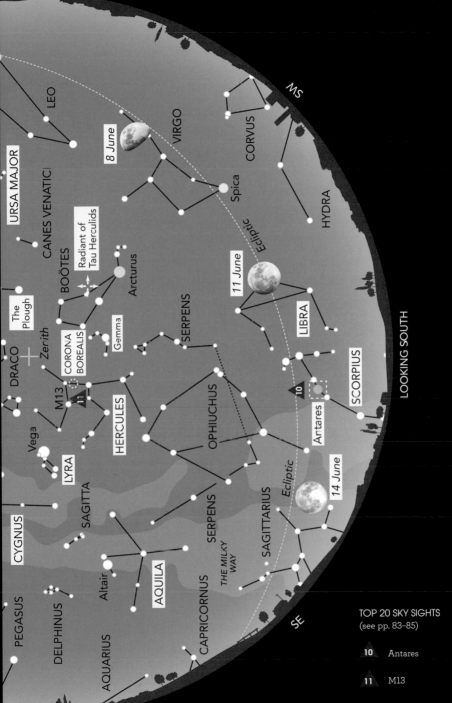

JUNE

MS

LEO

VIRGO

CORVUS

8 June

Spica

Ecliptic

HYDRA

URSA MAJOR

CANES VENATICI

Radiant of
Tau Herculids

BOÖTES

Arcturus

SERPENS

11 June

LIBRA

The
Plough

Zenith

Gemma

CORONA
BOREALIS

DRACO

SERPENS

SCORPIUS

M13

11

HERCULES

OPHIUCHUS

10

Antares

Vega

LYRA

Ecliptic

14 June

SAGITTA

SERPENS

SAGITTARIUS

CYGNUS

AQUILA

CAPRICORNUS

THE MILKY
WAY

SE

PEGASUS

Altair

DELPHINUS

AQUARIUS

LOOKING SOUTH

TOP 20 SKY SIGHTS
(see pp. 83–85)

10 Antares

11 M13

Though the sky never quite grows dark this month – especially for the northern regions of the British Isles – take advantage of the warm nights to enjoy the lovely summer constellations of **Hercules, Scorpius, Lyra, Cygnus** and **Aquila**. Stay up till the early hours, and you'll enjoy a procession of planets rising in the east.

JUNE'S CONSTELLATION

Ursa Major, the Great Bear, is an internationally favourite constellation. In Britain, its seven brightest stars are called the **Plough**, though children today often name it 'the saucepan'. In North America, it's known as the Big Dipper. Always on view in the northern hemisphere, the Plough is the first star pattern that most people get to know.

Look closely at the star in the middle of the bear's tail (or the handle of the saucepan), and you'll see it's double. **Mizar** and its fainter companion **Alcor** comprise one of the few binary stars you can 'split' with unaided eye.

And – unlike most constellations – the majority of the stars in the Plough lie at the same distance and were born together. Leaving aside **Dubhe** and **Alkaid**, the others are all moving in the same direction, along with other stars of the Ursa Major Moving Group, which include **Menkalinan** in **Auriga** and **Gemma** in **Corona Borealis** (see May's Constellation). Over thousands of years, the shape of the Plough will gradually change, as Dubhe and Alkaid go off on their own paths.

JUNE'S OBJECT

The name **Antares** means 'the rival of Mars' – and it's easy to see why. Look low in the south this month to spot this ruby jewel marking the heart of the constellation **Scorpius** (the Scorpion). Six hundred light years away, Antares is a bloated red giant star near the end of its life. Running low on its supplies of nuclear fuel, the star's core has shrunk while its outer layers have billowed out and cooled. Some 700 times wider than our star, Antares would engulf all the planets out to Mars if it replaced the Sun. Eventually, the core of Antares will collapse completely, and the star will explode as a brilliant supernova.

The giant star has a smaller companion (magnitude +5), which is hard to see against Antares's glare. Although it is a hot blue-white star, Antares B appears greenish in contrast to its bright red neighbour.

OBSERVING TIP

June is a great month for observing the Sun, with our local star at its highest in the sky. But be careful. NEVER look at the Sun directly, with your unprotected eyes or – especially – with a telescope or binoculars: it could blind you permanently. For naked-eye observing, use 'eclipse glasses' with filters (meeting the ISO 12312-2 safety standard) to reduce the Sun's radiation. You can project the Sun's image through binoculars or a telescope onto a piece of white card; or attach a filter made of special material (such as Baader AstroSolar® film) across the front of the instrument. Or go the whole hog: acquire solar binoculars or a solar telescope with built-in filters that you can use for viewing details of the Sun's churning surface.

JUNE'S TOPIC:
MOON ILLUSION

The Full Moon this month is bigger than usual, because it's a supermoon (see Special Events). And it will seem even more distended, thanks to the fact that the Full Moon in June hangs just above on the horizon: our companion world always looks larger when it is low down than when it's high in the sky.

This is called the 'Moon illusion', and – as the name suggests – it's not a physical effect but a trick played by the human eye. When you see the Moon behind a horizon of trees and houses, your mind compares it to these nearby objects, and so it appears larger. You can prove this by photographing the Moon at hourly intervals from rising to setting: you'll find no change in the size of the images.

You can make the illusion go away – at the risk of looking silly! – by standing with your back to the Moon, bending over and looking at it between your legs. The Moon will shrink to its normal size.

Ara is beneath the horizon as seen from the UK, and Peter Jenkins imaged the Fighting Dragons by remote observing, using an ASA 500-mm f/3.6 reflector located at El Sauce, Chile, hosted by Telescope Live. He acquired a total of 2 hours of exposures: 4 × 10 minutes each through filters passing light from hydrogen, oxygen and sulphur.

JUNE'S PICTURE

Below the tail of the celestial scorpion, **Scorpius**, lurks a pair of even more fearsome creatures: the Fighting Dragons of Ara. The lower dark region here is a dense cloud of gas and dust, 4000 light years away. At the top are massive hot stars, 'only' a few million years old, some of them over 30 times heavier than the Sun.

Radiation from these energetic stars is eating away the edge of the dark cloud, lighting it up as the bright nebula NGC 6188. The Fighting Dragons – spectacularly imaged here by Peter Jenkins – are tattered remnants of the dark cloud, silhouetted against this glowing gas. Eventually the dragons will both be slain, boiled away by the stars' radiation.

SUNDAY	MONDAY	TUESDAY	WEDNESDAY	THURSDAY	FRIDAY	SATURDAY
			1 Tau Herculids (am)	2 Moon near Castor and Pollux	3 Moon near Castor and Pollux	4
5 Moon near Regulus	6	7 3.48 pm First Quarter Moon	8	9 Moon near Spica	10 Moon near Spica	11
12 Venus near Uranus	13 Moon near Antares	14 12.52 pm Full Moon, supermoon	15	16 Mercury W elongation	17	18 Moon near Saturn (am)
19 Moon near Saturn (am)	20	21 4.11 am Last Quarter Moon near Jupiter (am); Summer Solstice	22 Moon between Jupiter and Mars (am)	23 Moon near Mars (am); Venus near the Pleiades (am)	24	25
26 Moon near Venus (am)	27 Moon near Mercury (am)	28	29 3.52 am New Moon	30		

SPECIAL EVENTS

- **Morning of 1 June:** Comet Schwassman-Wachman-3 may treat us to a spectacular storm of shooting stars, the **Tau Herculid meteor shower** (see May's Special Events).

- **14 June:** The second supermoon of 2022. It's only 1% smaller than next month's Full Moon, the year's best supermoon (see July's Special Events).

- **21 June, before dawn:** The bright 'star' near the Moon is Jupiter.

- **21 June, 10.13 am:** Summer Solstice. The Sun reaches its most northerly point in the sky, so today is Midsummer's Day, with the longest period of daylight and the shortest night.

- **22 June, before dawn:** You'll find Jupiter to the right of the Moon, and Mars to the left (Chart 6a).

- **23 June, before dawn:** Mars lies just to the right of the Moon.

- **23 June:** The BepiColombo probe will swing past Mercury for a second time; it will enter orbit about the innermost planet in 2025.

- **26 June, before dawn:** Venus and the crescent Moon make a stunning sight, with the Pleiades visible in binoculars above them and Mercury to the lower left (Chart 6b).

- **27 June, before dawn:** To the left of Venus, look for the thinnest crescent Moon very near the horizon, with Mercury below.

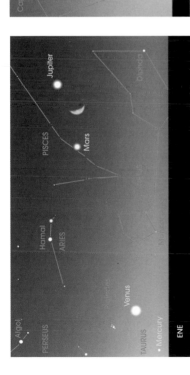

6a *22 June, 4 am. The Moon between Jupiter and Mars, with Venus.*

6b *26 June, 4 am. The crescent Moon close to Venus and the Pleiades.*

- There are no planets in the evening sky this month.
- **Saturn** is the first on the scene, rising in the south-east about 0.30 am and shining at magnitude +0.7 in Capricornus.
- Next to rise is faint **Neptune** (magnitude +7.9), which appears at around 1 am between Aquarius and Pisces.
- Mighty **Jupiter** rises in the east about 1.30 am,

at a brilliant magnitude –2.3 in Pisces.
- At the start of June, **Mars** is just to the lower left of Jupiter, and it moves steadily leftwards throughout the month. Lying in Pisces, the Red Planet shines at magnitude +0.6 and rises around 2 am.
- **Venus** surges above the north-eastern horizon about 3 am, at a magnificent magnitude –3.9. During the

month the Morning Star is tracking to the left against the background stars.
- On the morning of 12 June, Venus passes below **Uranus**, which is almost 10,000 times fainter at magnitude +5.8. The seventh planet lies in Aries, and also rises around 3 am. On 23 June, Venus passes under the Pleiades.
- You'll find **Mercury** to the lower left of Venus during the last week of June – the

period when the innermost planet is most easily visible this month. Mercury appears in the dawn twilight: low in the north-east, in the middle of June, about the time it reaches its greatest separation from the Sun (16 June). Gradually ascending in the sky, Mercury brightens throughout the month to reach magnitude –0.6 by the end of June, when it is rising at 3.30 am.

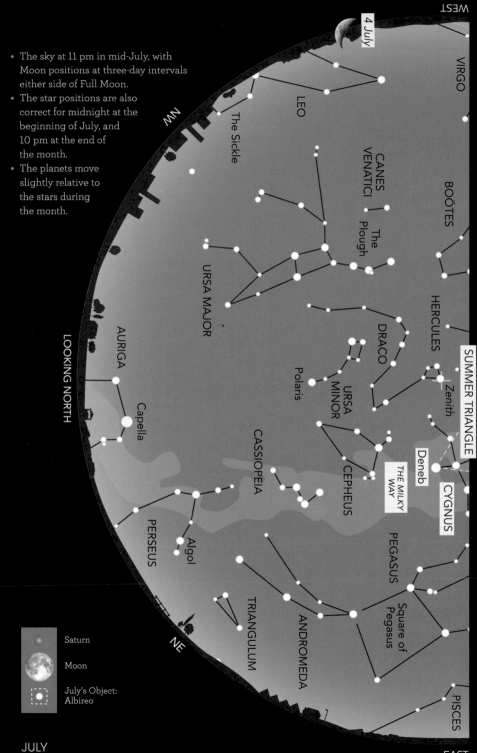

- The sky at 11 pm in mid-July, with Moon positions at three-day intervals either side of Full Moon.
- The star positions are also correct for midnight at the beginning of July, and 10 pm at the end of the month.
- The planets move slightly relative to the stars during the month.

WEST

4 July

VIRGO

NW

LEO

The Sickle

CANES VENATICI

The Plough

BOÖTES

HERCULES

URSA MAJOR

DRACO

SUMMER TRIANGLE

AURIGA

Polaris

URSA MINOR

Zenith

Capella

LOOKING NORTH

CASSIOPEIA

CEPHEUS

Deneb

THE MILKY WAY

CYGNUS

PERSEUS

Algol

PEGASUS

Saturn

Moon

July's Object: Albireo

TRIANGULUM

ANDROMEDA

Square of Pegasus

NE

PISCES

EAST

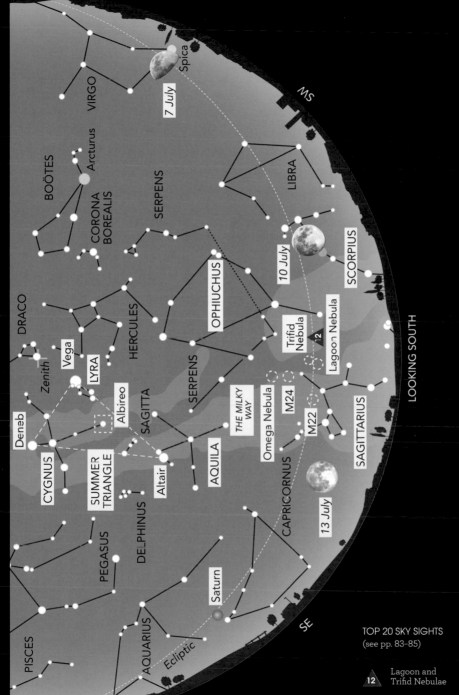

JULY

MS

LOOKING SOUTH

SE

VIRGO

Spica

7 July

Arcturus

BOÖTES

CORONA BOREALIS

SERPENS

LIBRA

10 July

SCORPIUS

DRACO

Vega

LYRA

Zenith

Albireo

HERCULES

OPHIUCHUS

Trifid Nebula

Lagoon Nebula

12

Deneb

SAGITTA

SERPENS

THE MILKY WAY

Omega Nebula

M24

M22

SAGITTARIUS

CYGNUS

SUMMER TRIANGLE

Altair

AQUILA

CAPRICORNUS

13 July

DELPHINUS

PEGASUS

Saturn

PISCES

AQUARIUS

Ecliptic

TOP 20 SKY SIGHTS
(see pp. 83–85)

12   Lagoon and
Trifid Nebulae

JULY   **45**

Two of the most ancient constellations, **Sagittarius** and **Scorpius**, are at their best this month, decorating the southern region of the sky. Their stars lie in front of the brightest parts of the **Milky Way**, towards the heart of our Galaxy. Higher in the sky, the prominent **Summer Triangle** is composed of **Vega**, **Deneb** and **Altair**, the leading lights of **Lyra**, **Cygnus** and **Aquila**.

JULY'S CONSTELLATION

Low in the south this month you'll find a constellation that's shaped rather like a teapot, with the handle to the left and the spout to the right.

To the ancient Greeks, the star pattern **Sagittarius** resembled an archer, with the torso of a man and the body of a horse. The 'handle' of the teapot represents his upper body, while the 'spout' depicts his bow and an arrow aimed at **Scorpius**, the celestial scorpion.

Sagittarius is rich in nebulae and star clusters, easily visible in binoculars.

Above the spout lies the wonderful **Lagoon Nebula** – a region of starbirth that's visible to the naked eye on a really dark night. Its neighbour, the three-lobed **Trifid Nebula**, requires a telescope (see August's Objects).

Between Sagittarius and **Aquila**, you'll find a bright patch of stars in the Milky Way, catalogued as **M24**. Raise your binoculars higher to spot another star-forming region, the **Omega Nebula**. Finally, the fuzzy patch **M22** is a globular cluster of almost a million stars, lying 11,000 light years away.

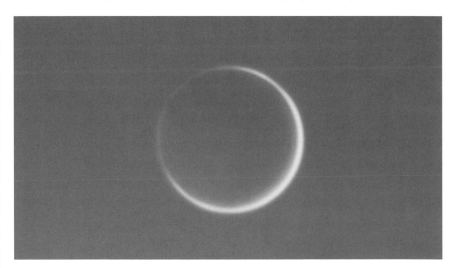

On the morning of 2 June 2020, Pete Lawrence recorded 10,098 frames of Venus using a ZWO ASI174MM planetary camera fitted with a 742-nm infrared pass filter (to increase contrast), all attached to a 100-mm refractor. He obtained the sky colour from a DSLR camera set up in parallel.

JULY'S OBJECT

Albireo marks the head of the constellation **Cygnus**, a swan with outspread wings flying down the **Milky Way**. Though it may look unassuming to the unaided eye, turn even the smallest telescope on Albireo and you'll find it's a double star – and one of the most glorious sights in the night sky: a dazzling yellow star teamed up with a blue companion.

The yellow star is a giant, near the end of its life. It's 60 times bigger than the Sun, and 1200 times brighter. The fainter blue companion is 'only' 200 times brighter than the Sun.

The spectacular colour contrast is due to the stars' different temperatures. The giant star is slightly cooler than our Sun, and has a similar hue. The smaller companion is far hotter: it's so incandescent that it shines not merely white-hot, but blue-white.

JULY'S TOPIC:
THE 13TH ZODIAC SIGN

On its monthly journey round the Earth, the Moon spends 10–11 July in the constellation of **Ophiuchus**. But hang on, I hear you saying: surely the Moon only travels though the well-known constellations of the Zodiac, like Leo, Gemini or Taurus?

That's not entirely true. The ancient astronomers of Mesopotamia noticed that the Sun, Moon and planets kept to a distinct band in the sky, which they divided into distinctive star patterns, or 'signs'. To the Greeks, these constellations largely depicted creatures – such as a crab, a pair of fish and even a sea goat – and so they named it the Zodiac, after their word for 'animal'. Unfortunately, there were 13 star patterns in the Zodiac, and that didn't fit in with the 12 months

OBSERVING TIP

This is the month when you really need a good, unobstructed horizon to the south, for the best views of the glorious summer constellations of Scorpius and Sagittarius. They never rise high in temperate latitudes, so make the best of a southerly view – especially over the sea – if you're away on holiday. A good southern horizon is also best for views of the planets, because they rise highest when they're in the south.

of the year. Dull sprawling Ophiuchus (the Serpent Bearer) was the one to get the chop, in favour of its showier neighbour Scorpius (the Scorpion).

Every few years, newspaper headlines shout 'Astronomers have discovered a new sign of the Zodiac!' But the 13th zodiacal constellation actually dates back thousands of years.

JULY'S PICTURE

This is Venus – but in a most unusual light. Sunlight is shining through the planet's atmosphere, to create an ethereal halo around the silhouette of Venus itself.

Once every 19 months, Venus passes between the Earth and the Sun. Occasionally, they are exactly in line and we see Venus in transit across the solar disc. More often, the inner planet is above or below the Sun, so close to the brilliant solar orb that it's difficult to observe Venus at all.

It needed very carefully planning (and extra care to ensure his telescope didn't accidentally point at the Sun) for Pete Lawrence to take this rare image of Venus's atmospheric ring in June 2020 – a feat that's been described as requiring the equivalent of a black belt in astrophotography!

SUNDAY	MONDAY	TUESDAY	WEDNESDAY	THURSDAY	FRIDAY	SATURDAY
31					1	2 Moon near Regulus
3 Moon near Regulus	4 Earth at aphelion	5	6	7 3.14 am First Quarter Moon near Spica	8	9
10 Moon near Antares	11	12	13 7.37 pm Full Moon, supermoon	14	15 Moon near Saturn	16
17	18	19 Moon near Jupiter (am)	20 3.18 pm Last Quarter Moon	21 Moon near Mars (am)	22 Moon near Mars (am)	23
24	25	26 Moon near Venus (am)	27 Moon near Venus (am)	28 6.55 pm New Moon	29	30

Supermoon

SPECIAL EVENTS

- **4 July, 8.11 am:** The Earth is at its furthest from the Sun (aphelion), just over 152 million km away.
- **13 July:** The brightest Moon of 2022. This year there are four supermoons – Full Moons within 367,607 km of the Earth – between May and August, but this month's is the closest, with the Moon just 357,418 km distant. It's 30% brighter than the faintest Full Moon.
- **15 July:** Saturn lies directly above the Moon.
- **19 July, am:** The brilliant 'star' by the Moon is giant planet Jupiter (Chart 7a).
- **21–22 July, before dawn:** The Moon is near Mars.
- **26–27 July, before dawn:** Low in the north-east, the crescent Moon forms a striking pair with Venus (Chart 7b).

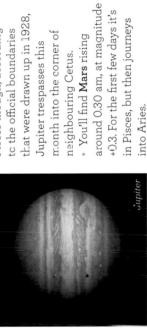

7a 19 July, 1 am. Jupiter lies above the Moon.

7b 26–27 July, 3.45 am. The crescent Moon with Venus.

- **Saturn** is leading a veritable parade of planets that's starting to rise at around midnight. The ringworld appears above the horizon in the south-east at about 10.30 pm: it lies in Capricornus and shines at magnitude +0.5.

- Dim **Neptune** (magnitude +7.9) follows Saturn at around 11.30 pm; it can be seen on the borders of Aquarius and Pisces.

- Next comes the brilliant **Jupiter**, blazing at magnitude –2.6. The giant planet rises

Jupiter

at about midnight. It lies within the outline of the ancient star pattern of Pisces though, according to the official boundaries that were drawn up in 1928, Jupiter trespasses this month into the corner of neighbouring Cetus.

- You'll find **Mars** rising around 0.30 am, at magnitude +0.3. For the first few days it's in Pisces, but then journeys into Aries.

- **Uranus** resides in Aries all month. It rises about 1 am, shining at a feeble magnitude +5.8. By the end of the month, Mars has moved close to the seventh planet.

- **Venus** is a glorious Morning Star all month, rising around 3 am, its radiance putting all the stars and other planets to shame at a magnificent magnitude –3.9.

- **Mercury** is too close to the Sun to be visible this month.

JULY'S PLANET WATCH

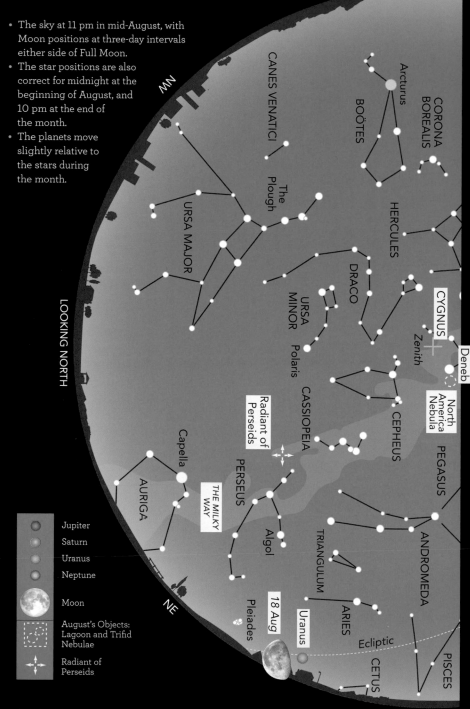

- The sky at 11 pm in mid-August, with Moon positions at three-day intervals either side of Full Moon.
- The star positions are also correct for midnight at the beginning of August, and 10 pm at the end of the month.
- The planets move slightly relative to the stars during the month.

WEST

LOOKING NORTH

NW

NE

EAST

CANES VENATICI

BOÖTES

Arcturus

CORONA BOREALIS

HERCULES

The Plough

URSA MAJOR

DRACO

URSA MINOR

Polaris

CYGNUS

Zenith

Deneb

North America Nebula

CASSIOPEIA

CEPHEUS

PEGASUS

Radiant of Perseids

Capella

PERSEUS

AURIGA

THE MILKY WAY

Algol

TRIANGULUM

ANDROMEDA

Pleiades

18 Aug

Uranus

ARIES

Ecliptic

CETUS

PISCES

Jupiter
Saturn
Uranus
Neptune
Moon
August's Objects: Lagoon and Trifid Nebulae
Radiant of Perseids

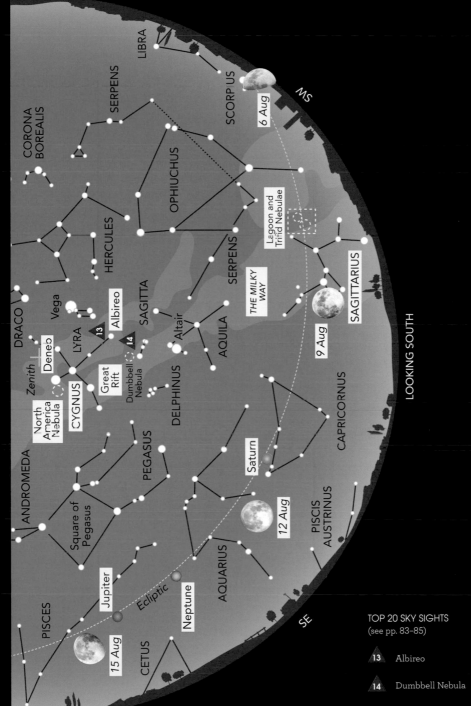

WEST

LIBRA
SERPENS
CORONA BOREALIS
SCORPIUS
6 Aug
MS
OPHIUCHUS
HERCULES
SERPENS
DRACO
Vega
Albireo
SAGITTA
Lagoon and Trifid Nebulae
THE MILKY WAY
SAGITTARIUS
9 Aug
LYRA
Deneb
Zenith
13
14
Altair
AQUILA
CYGNUS
Great Rift
Dumbbell Nebula
North America Nebula
DELPHINUS
CAPRICORNUS
LOOKING SOUTH
ANDROMEDA
PEGASUS
Saturn
12 Aug
PISCIS AUSTRINUS
Square of Pegasus
Jupiter
AQUARIUS
PISCES
Ecliptic
Neptune
15 Aug
CETUS
SE

AUGUST

TOP 20 SKY SIGHTS
(see pp. 83–85)

13 Albireo

14 Dumbbell Nebula

At last we have planets in the evening sky! There's been a dearth of the Solar System's brighter worlds since February, unless you are an early bird. But now we're treated to a procession of planets that will brighten up our skies until the end of the year, led by **Saturn**, which is at its closest to Earth this month.

AUGUST'S CONSTELLATION

The flying swan – **Cygnus** – actually looks like its namesake, with outspread wings and an elongated neck. According to myth, it commemorates the occasion when Zeus disguised himself as a swan to seduce Leda, the wife of King Tyndareus of Sparta. She gave birth to two eggs, containing mortal and immortal offspring: they included Castor and Pollux (see March's Constellation) and the legendary beauty Helen of Troy.

Deneb forms the tail of the swan. It's the most distant of the first-magnitude stars, but its distance is hard to pin down. Estimates range from 1500 to 2600 light years: which means that the star shines anything between 50,000 and 200,000 times brighter than the Sun.

The swan's head is marked by **Albireo** (see July's Object), probably the most beautiful double star in the sky. Another Cygnus gem is the **North America Nebula**, a cloud of gas larger than the Full Moon; on a dark night you can pick it out with binoculars. The **Milky Way** meanders through Cygnus, split by the dark silhouette of the **Great Rift**.

AUGUST'S OBJECTS

A pair of glorious nebulae adorn the constellation **Sagittarius**. The brighter is the **Lagoon Nebula**, just visible to the unaided eye as a glowing cloud three times the width of the Moon. Lying about 5000 light years away, the Lagoon is illuminated by a newly born star that's 40 times heavier than the Sun and 200,000 times brighter. Dotted over the glowing lagoon are small dark clouds, each poised to condense into new stars and planets.

Its companion, the **Trifid Nebula**, looks dramatically different. It appears split into three lobes (the meaning of the word 'trifid') by bands of dark cosmic dust silhouetted against its bright core. The nebula is packed with young stars and embryonic stars that have yet to fully form.

AUGUST'S TOPIC:
CENTRE OF THE MILKY WAY

The **Milky Way**, stretching across the sky as a gently glowing band, is our inside view of a flattened galaxy consisting of hundreds of billions of stars. We live about halfway out, and from our viewpoint the centre of the Milky Way lies in the direction of **Sagittarius**.

OBSERVING TIP

When you first go out to observe, you may be disappointed at how few stars you can see in the sky. But wait for 20 minutes, and you'll be amazed at how your night vision improves. One reason for this 'dark adaption' is that the pupil of your eye grows larger. More importantly, in dark conditions the retina of your eye builds up bigger reserves of rhodopsin, the chemical that responds to light.

The Great Conjunction: setting my Samsung Galaxy S10e to automatic, the exposure was 0.25 second at f/1.5 and ISO 1250.

Unfortunately, whatever happens downtown in our Galaxy is hidden – even for the most powerful optical telescopes – by dense clouds of dark dust. But telescopes observing infrared and radio waves have lifted the veil on the Galaxy's heart, and revealed a maelstrom of activity.

Vast clouds of gas are laced with powerful magnetic fields, channelling streams of speeding subatomic particles. Towards the centre, stars are ever more closely packed and fast moving, with speeds up to 18 million kilometres per hour.

The beating heart of our Galaxy is marked by a compact but very energetic object, Sagittarius A* (pronounced 'A-star'), containing a supermassive black hole as heavy as 4 million Suns. Infalling gas on the brink of the black hole emits powerful bursts of radiation, just before it disappears from our Universe.

But don't worry about this monster unleashing devastation on our planet: Sagittarius A* lies at a safe distance of 26,700 light years away.

AUGUST'S PICTURE

Every cloud has its silver lining, they say, and when there are no clouds it's even better! In December 2020, COVID-19 restrictions meant I couldn't fly directly from the UK to the USA, so I sat out my quarantine in Barbados. At the time, astronomers around the world were waiting for the Great Conjunction of Jupiter and Saturn (see February's Picture), often under frustrating cloud cover.

On 16 December, a few days before the conjunction, the sky was brilliantly clear from my hotel terrace, revealing a stunning tableau of the encroaching planets along with the crescent Moon. Travelling light, I didn't have my telescope or DSLR, with me, but my smartphone held steadily against the railing recorded the sight for posterity.

SUNDAY	MONDAY	TUESDAY	WEDNESDAY	THURSDAY	FRIDAY	SATURDAY
	1 Mars near Uranus	2	3 Moon near Spica	4	5 12.06 pm First Quarter Moon	6 Moon occults Dschubba
7	8	9	10	11 Moon near Saturn	12 2.36 am Full Moon, supermoon; Perseids	13 Perseids (am)
14 Saturn opposition; Moon near Jupiter	15 Moon near Jupiter	16	17	18	19 5.36 am Last Quarter Moon near Mars and the Pleiades (am)	20 Moon near Mars, the Pleiades and Aldebaran (am)
21 Mars near the Pleiades (am)	22	23	24 Moon near Castor and Pollux (am)	25	26 Moon near Venus	27 9.17 am New Moon; Mercury E elongation
28	29	30 Moon near Spica	31 Moon near Spica			

SPECIAL EVENTS

- **6 August, 10.40–11.10 pm:** The Moon moves in front of Dschubba (magnitude +2.3) in Scorpius (the timing will vary by a few minutes depending on your location).
- **11 August:** Saturn lies to the upper left of the Moon.
- **Night of 12/13 August:** Maximum of the **Perseid meteor shower**. Sadly, one of the best annual shooting star displays will be washed out by light pollution from the Full Moon, especially as it's a supermoon (see July's Special Events).
- **14 August:** Saturn is opposite to the Sun in the sky, and closest to the Earth at 1325 million km (see Planet Watch).
- **14–15 August:** Brilliant Jupiter pairs up with the Moon.
- **19–20 August, am:** The Moon passes Mars, the Pleiades and Aldebaran (Chart 8a).
- **21 August, am:** Mars lies directly below the Pleiades, with Aldebaran and the Moon to the left.
- **26 August, before dawn:** On the eastern horizon, you may just catch the thinnest crescent Moon lying to the left of Venus (Chart 8b).
- This month, NASA plans to launch a spacecraft to investigate asteroid Psyche, a lump of iron 200 km wide.
- The European-Russian mission Exomars is scheduled to head for Mars: it comprises the Kazachok lander and the Rosalind Franklin rover that will search for life.

8a 19–20 August, 2 am. The Moon passes Mars, the Pleiades and Aldebaran.

8b 26 August, 5 am. The crescent Moon lies close to Venus.

• **Saturn** lies in Capricornus and is visible all night long. The sixth planet reaches its brightest this year (magnitude +0.3) when it is opposite the Sun and nearest the Earth on 14 August. Through a small telescope, astound yourself by viewing Saturn's breathtaking rings.

• **Neptune**, on the borders of Aquarius and Pisces, glows at a mere magnitude +7.8, rising around 9.30 pm.

• Clearing the horizon about 10 pm, **Jupiter** blazes at magnitude –2.8. This month, it has strayed slightly out of the traditional Zodiac constellation of Pisces into neighbouring Cetus.

• **Uranus** (magnitude +5.8) rises around 11 pm in Aries.

• On 1 August, **Mars** also lies in Aries, directly below Uranus and outshining the distant world a hundred times over, at magnitude +0.1.

Rising about 11.30 pm, the Red Planet moves inexorably leftwards during the month into Taurus. It passes below the Pleiades on 20 and 21 August, with the waning Moon nearby (Chart 8a). Towards the end of August, Mars approaches Aldebaran.

• Rising around 4 am, **Venus** is incomparable in brightness at magnitude –3.9. During August, the Morning Star is gradually sinking down into the morning twilight glow.

• **Mercury** is lost in the Sun's glare this month, even when it reaches greatest separation on 27 August.

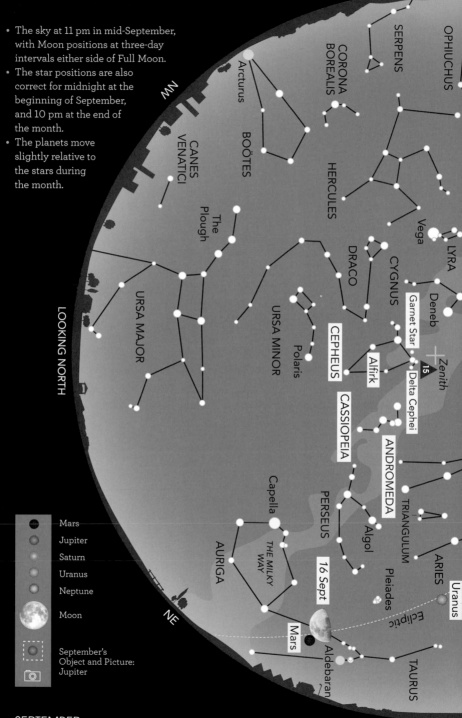

- The sky at 11 pm in mid-September, with Moon positions at three-day intervals either side of Full Moon.
- The star positions are also correct for midnight at the beginning of September, and 10 pm at the end of the month.
- The planets move slightly relative to the stars during the month.

WEST

LOOKING NORTH

NW

NE

EAST

OPHIUCHUS
SERPENS
CORONA BOREALIS
Arcturus
CANES VENATICI
BOÖTES
HERCULES
The Plough
Vega
LYRA
DRACO
CYGNUS
Deneb
Garnet Star
Zenith
15
Delta Cephei
URSA MAJOR
URSA MINOR
Polaris
CEPHEUS
Alfirk
CASSIOPEIA
ANDROMEDA
TRIANGULUM
PERSEUS
Capella
Algol
ARIES
Uranus
AURIGA
THE MILKY WAY
Pleiades
16 Sept
Mars
Aldebaran
TAURUS
Ecliptic

Mars
Jupiter
Saturn
Uranus
Neptune
Moon
September's Object and Picture: Jupiter

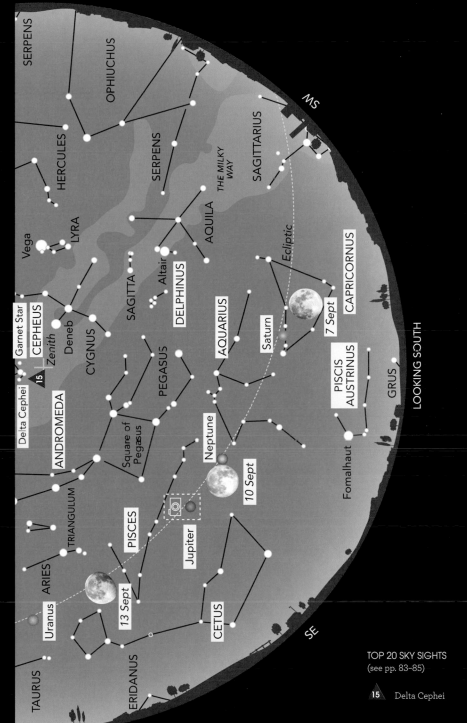

SEPTEMBER

WEST

SERPENS

OPHIUCHUS

HERCULES

LYRA

Vega

SAGITTA

SERPENS

AQUILA

THE MILKY WAY

SAGITTARIUS

MS

Ecliptic

Garnet Star
CEPHEUS
Zenith
Delta Cephei
15
Deneb
CYGNUS

ANDROMEDA

Altair

DELPHINUS

SAGITTA

AQUARIUS

Saturn

7 Sept

CAPRICORNUS

LOOKING SOUTH

PEGASUS

Square of Pegasus

Neptune

10 Sept

PISCIS AUSTRINUS

GRUS

TRIANGULUM

PISCES

Jupiter

Fomalhaut

ARIES

Uranus

13 Sept

CETUS

SE

TAURUS

ERIDANUS

TOP 20 SKY SIGHTS
(see pp. 83–85)

15 Delta Cephei

EAST

SEPTEMBER

Five planets are strutting their stuff this month: **Jupiter** is at its most brilliant, while dim **Uranus** pulls off the trick of disappearing behind the Moon. They swim in the sky among the watery constellations of autumn. **Aquarius** (the Water Carrier) pours a stream of fluid down to **Piscis Austrinus** (the Southern Fish), surrounded by **Delphinus** (the Dolphin), the strangely named Sea Goat (**Capricornus**), a pair of Fishes (**Pisces**) and **Cetus** (the Sea Monster).

SEPTEMBER'S CONSTELLATION

Shaped like a child's drawing of a house, **Cepheus** represents a mythical king of Ethiopia. He was married to Queen **Cassiopeia** who – both in legend and as a constellation – is far more exciting (she once boasted that her daughter **Andromeda** was more beautiful than the sea nymphs, with almost disastrous effects).

Three fascinating stars lift Cepheus from near obscurity. **Alfirk** is a double star, its companion being visible through a small telescope. The aptly named **Garnet Star** – named by William Herschel because of its strong ruddy hue – is a hypergiant, some 1000 times wider than the Sun, and varies in brightness between magnitudes +3.4 and +5.1 over a couple of years. The star of the show, though, is

OBSERVING TIP

It's best to view your favourite objects when they're well clear of the horizon. If you observe them low down, you're looking through a large thickness of the atmosphere – which is always shifting and turbulent. It's like trying to observe the outside world from the bottom of a swimming pool! This turbulence makes the stars appear to twinkle. Low-down planets also twinkle – although to a lesser extent, because they subtend tiny discs, and aren't so affected.

Delta Cephei. As this star pulsates in size, its brightness changes from magnitude +3.5 to +4.4 over a period of five days and nine hours. Astronomers have found that the period of variation of stars like this (Cepheid variables) is closely related to their intrinsic brilliance, allowing them to be used as pulsating stellar beacons to measure cosmic distances.

SEPTEMBER'S OBJECT

Almost 143,000 kilometres in diameter, **Jupiter** is the biggest world in our Solar System, and could contain 1300 planets the size of ours. The cloudy gas giant is also very efficient at reflecting sunlight: at its closest to the Earth this month, Jupiter shines at a dazzling magnitude –2.9. It's a fantastic target for stargazers, whether you're using your unaided eyes, binoculars or a small telescope.

Despite its size, Jupiter spins faster than any other planet, in only 9 hours 55 minutes. As a result, its equator bulges outwards. Through a small telescope you can make out bright and dark bands stretched round the tangerine-shaped planet. These are white clouds of frozen ammonia overlying the planet's darker lower atmosphere.

Jupiter commands a family of about 80 moons. The four biggest are visible in binoculars, and even – to the really sharp-sighted – with the unaided eye. Ganymede is larger than the planet Mercury, while Callisto has the oldest unchanged surface of any world. Io sports hundreds of active volcanoes. Europa's icy surface conceals a deep ocean which may be home to some kind of aquatic life.

SEPTEMBER'S TOPIC: THE EQUINOX

On 23 September, the rising Sun shines directly down a long stone-lined passageway into the heart of an ancient burial mound called La Hougue Bie, on the island of Jersey. It marks a special date in the year: the Autumn Equinox, when day and night are the same length.

Other ancient sites, like Stonehenge and Newgrange in Ireland, are lined up with the rising or setting of the Sun on the shortest day, the Winter Solstice in December. Taken together, these monuments show our distant ancestors were well versed in astronomy and its link to the seasons.

In the northern hemisphere, the Sun is highest during the months from March to September, its heat shining down more directly on the ground and bringing the warmth of summer. Conversely, the weather is coldest between September and March. And the balance point comes with the equinoxes – a word meaning 'equal night'.

SEPTEMBER'S PICTURE

Jupiter's Great Red Spot is a perennial favourite for astrophotographers. But not many have viewed it from such an unusual perspective as Pete Williamson in Shropshire – observing with NASA's Juno spacecraft (see page 89).

Jupiter's winds organise its clouds into elongated bands, wracked by myriad storms as we can see in this close-up. The greatest weather system is the Great Red Spot, larger than the Earth. Ultraviolet radiation from the Sun converts the gas in its upper regions into complex colourful molecules.

Juno snapped this image as it swept over Jupiter on 2 June 2020, and Pete Williamson downloaded it from the JunoCam website as a mapprojected.png file. He adjusted the colours, image depth, brightness and saturation levels using the free software GIMP.

SEPTEMBER'S CALENDAR

SUNDAY	MONDAY	TUESDAY	WEDNESDAY	THURSDAY	FRIDAY	SATURDAY
				1	2	3 7.08 pm First Quarter Moon near Antares
4	5	6	7	8 Moon near Saturn	9	10 10.59 am Full Moon
11 Moon near Jupiter	12	13	14 Moon occults Uranus	15 Moon near the Pleiades	16 Neptune opposition; Moon near Mars and Aldebaran	17 10.52 pm Last Quarter Moon
18	19	20 Moon near Castor and Pollux (am)	21	22	23 Autumn Equinox	24
25 10.54 pm New Moon	26 Jupiter opposition	27	28	29	30 Moon near Antares	

SPECIAL EVENTS

- **8 September:** Saturn lies to the upper right of the Moon.
- **11 September:** The Moon and Jupiter shine close together in the south-east.
- **14 September, 10.27–11.20 pm:** The Moon moves in front of Uranus: the timing will vary by a few minutes, depending on your location (Chart 9a).

Such events are very rare, with the last occultation of Uranus visible from the British Isles being back in 1953. But, amazingly, there are two Uranus occultations in 2022 (see December's Special Events).

- **15 September:** The Moon lies below the Pleiades (Chart 9b).

- **16 September:** Neptune is opposite to the Sun, and at is nearest to the Earth at 4325 million km (see Planet Watch).
- **16 September:** The Moon passes just above Mars, with Aldebaran to the lower right and the Pleiades to the upper right (Chart 9b).

- **23 September, 2.04 am:** Nights become longer than days as the Sun moves south of the Equator at the Autumn Equinox.
- **26 September:** Jupiter is at its closest to the Earth this year, 591 million km away, and opposite to the Sun in the sky (see Planet Watch).

Uranus, disappearance

Uranus, reappearance

9a 14 September, 10.27–11.20 pm. The Moon occults Uranus.

Capella
AURIGA
PERSEUS
The Pleiades
CETUS
Menkar
El Nath
16 Sept
15 Sept
Mars
TAURUS
Aldebaran
The Hyades
NE
E

9b 15–16 September, 11 pm. The Moon passes the Pleiades and Mars.

• Giant planet **Jupiter** is the 'star' of the month, closest to the Earth on 26 September when it is also at its brightest this year, at a radiant magnitude –2.9. Jupiter lies in Pisces, and is visible all night long. With binoculars or a small telescope, seek out its four largest moons, constantly changing position each night as they circle the mighty world.

• Well to the right of Jupiter, **Saturn** lies in Capricornus, setting around 3.30 am.

At magnitude +0.4, it's 20 times fainter than Jupiter.

• Between the Solar System's two giants, **Neptune** is lurking in Aquarius. Like Jupiter, the most distant planet is at its closest to us this month, on 16 September, and is above the horizon all night. But even at opposition, Neptune only reaches magnitude +7.8, so you'll need binoculars or a telescope to see it at all.

• **Uranus** rises about 9 pm, in Aries. At magnitude +5.7, it's just visible to the naked eye, but much better seen with optical aid. And be sure to whip out your binoculars or telescope on the evening of 14 September, when Uranus suffers a very rare occultation by the Moon (see Special Events and Chart 9a).

• **Mars** lies in Taurus, starting the month below the Pleiades and above Aldebaran. Try comparing the colour of the Red Planet with the hue of the slightly fainter red giant star, preferably using binoculars. Mars rises around 10 pm and shines at magnitude –0.3.

• About 5.30 am you'll find **Venus** rising in the eastern twilight. The glorious Morning Star is shining at magnitude –3.9, and during the month it sinks even lower into the glow of dawn, to disappear by the end of September.

• **Mercury** is too close to the Sun to be seen this month.

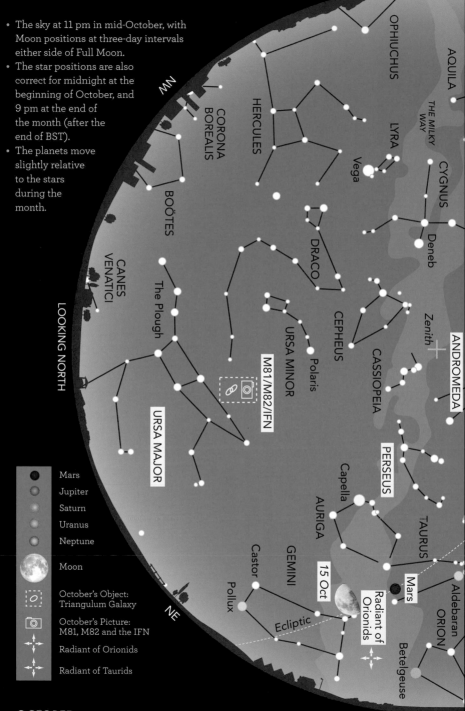

- The sky at 11 pm in mid-October, with Moon positions at three-day intervals either side of Full Moon.
- The star positions are also correct for midnight at the beginning of October, and 9 pm at the end of the month (after the end of BST).
- The planets move slightly relative to the stars during the month.

WEST

OPHIUCHUS

AQUILA

THE MILKY WAY

CORONA BOREALIS

HERCULES

LYRA

CYGNUS

Vega

NW

Deneb

BOÖTES

DRACO

CEPHEUS

Zenith

ANDROMEDA

CANES VENATICI

The Plough

URSA MINOR

Polaris

CASSIOPEIA

M81/M82/IFN

LOOKING NORTH

URSA MAJOR

PERSEUS

Capella

AURIGA

TAURUS

Castor

GEMINI

15 Oct

Mars

Aldebaran

Pollux

Radiant of Orionids

ORION

NE

Ecliptic

Betelgeuse

Mars	
Jupiter	
Saturn	
Uranus	
Neptune	
Moon	
October's Object: Triangulum Galaxy	
October's Picture: M81, M82 and the IFN	
Radiant of Orionids	
Radiant of Taurids	

EAST

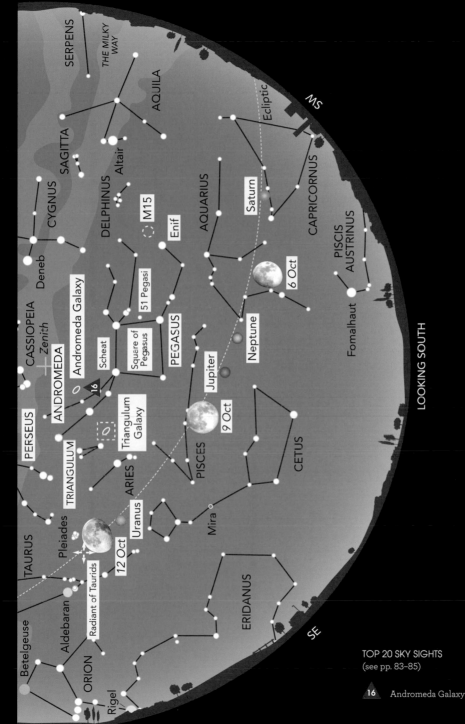

WEST

SERPENS

THE MILKY WAY

AQUILA

SAGITTA

CYGNUS

Altair

DELPHINUS

Deneb

CASSIOPEIA

Zenith

ANDROMEDA

Andromeda Galaxy

Scheat

Square of Pegasus

PEGASUS

51 Pegasi

M15

Enif

AQUARIUS

Saturn

Ecliptic

SW

CAPRICORNUS

PISCIS AUSTRINUS

6 Oct

Neptune

Fomalhaut

Jupiter

9 Oct

PERSEUS

TRIANGULUM

Triangulum Galaxy

16

ARIES

Uranus

PISCES

CETUS

Mira

TAURUS

Pleiades

12 Oct

Aldebaran

Radiant of Taurids

ERIDANUS

SE

Betelgeuse

ORION

Rigel

LOOKING SOUTH

OCTOBER

TOP 20 SKY SIGHTS
(see pp. 83–85)

16   Andromeda Galaxy

EAST

Look out for fireworks, of the celestial kind, this month. On 21 October, debris from Halley's Comet will light up the sky, while on the last couple of nights of the month we may be treated to an unusual display of the **Taurid** shooting stars. The planets are well on view in the evening sky, and there's a partial eclipse of the Sun on 25 October.

OCTOBER'S CONSTELLATION

Though **Pegasus** appears as little more than a large, empty square of four medium-bright stars, our ancestors managed to see it as an upside-down winged horse. In legend, Pegasus sprang from the blood of Medusa the Gorgon when **Perseus** severed her head.

A red giant star 100 times wider than the Sun, **Scheat** is nearing the end of its life and varies in brightness slightly as it pulsates. **Enif** (the nose) is an orange supergiant, with a faint blue companion visible in a small telescope, or even good binoculars.

Next to Enif – and Pegasus's best-kept secret – is the beautiful globular cluster

---

OBSERVING TIP

The Andromeda Galaxy is often described as the furthest object 'easily visible to the unaided eye'. It can be a bit elusive, though – especially if you are suffering from light pollution. The trick is to memorise Andromeda's pattern of stars, and then to look slightly to the *side* of where you expect the galaxy to be. This technique – called 'averted vision' – causes the image to fall on the outer region of your retina, which is more sensitive to light than the central region that's evolved to discern fine details. The technique is also crucial when you want to observe the faintest nebulae or galaxies through a telescope.

M15. You'll need a telescope for this one. M15 is 33,000 light years away, and contains over 100,000 densely packed stars.

And Pegasus contains the first planet to be discovered beyond our Solar System, orbiting the star **51 Pegasi**, which is just visible to the unaided eye. The planet has been named Dimidium (meaning 'half' in Latin, as it's half the mass of Jupiter).

OCTOBER'S OBJECT

This month, we have a little-known extragalactic neighbour, the **Triangulum Galaxy**. To find the elusive beast, start with the great **Andromeda Galaxy** and then – preferably with binoculars or a low-powered telescope – move the same distance below the line of stars making up the constellation of **Andromeda**. In the faint but distinctive star pattern of **Triangulum** (the Triangle), you should hit another fuzzy patch of light. Under exceptional conditions, the Triangulum Galaxy is *just* visible to the unaided eye (although it helps to be in a desert!).

But it's a challenge to see any detail in this scruffy little spiral even through a moderate telescope. It has a very low surface brightness, and – at 3 million light years' distance – it's slightly further away than the Andromeda Galaxy.

The Triangulum Galaxy (catalogued as M33) is the third-largest galaxy in the

It took a total of 25 hours' observing for Simon Hudson to obtain this deep image, using a 70-mm refractor (Altair Astro 70 EDQ-R) with a QHY9S CCD camera, some exposures through red, green and blue Baader filters (see page 88 for more details).

Local Group, and is home to 40 billion stars: about one-tenth the population of our Milky Way. But it's a hotbed of star formation, boasting an enormous star-forming nebula, NGC 604, which measures 1500 light years across.

OCTOBER'S TOPIC: JAMES WEBB SPACE TELESCOPE

A hundred times more powerful than the Hubble Space Telescope, a new 'eye on the sky' promises to revolutionise our knowledge of the Cosmos.

This international telescope is named after NASA Administrator James Webb, who – in the 1960s – insisted that the agency prioritised space science and missions to the planets.

Like Hubble, the James Webb Space Telescope observes from above the Earth's churning atmosphere. But it's placed much further from our planet, for a more unobstructed view. Webb has a main mirror almost three times larger than Hubble's: too wide to be launched on any rocket, this mirror is made of 18 segments that are folded up for launch and then unfurled. Designed to detect the faintest infrared radiation from space, Webb's telescope is protected from the Sun's warmth by a giant heat-shield the size of a tennis court.

With its penetrating stare, Webb will lay bare the birth of the first galaxies, reveal how stars and their planets are created, and check out distant worlds where alien life may exist.

OCTOBER'S PICTURE

There's a familiar pair of galaxies in this image – **M82** (above) and **M81** in **Ursa Major** – but you might be wondering why Simon Hudson took this picture on a cloudy night. In fact, there would be little point in waiting for these shining wisps to disappear: they lie thousands of light years away in our Galaxy.

The whole of interstellar space is filled with dusty clouds of gas, but we don't usually see them unless they're illuminated by a nearby star, or are so dense as to be opaque. In 1983, the pioneering Infrared Astronomical Satellite discovered radiation from warm dust scattered all over the sky. Long-exposure optical exposures then revealed these interstellar clouds faintly illuminated by the combined light of the stars in the Milky Way, making them 'integrated flux nebulae' (**IFN**).

Personally, though, I prefer an earlier term that better encapsulates their ethereal beauty: galactic cirrus.

SUNDAY	MONDAY	TUESDAY	WEDNESDAY	THURSDAY	FRIDAY	SATURDAY
30 British Summer Time ends; Taurids	31 Taurids					1
2	3 1.14 am First Quarter Moon	4	5 Moon near Saturn	6	7	8 Mercury W elongation; Moon near Jupiter
9 9.55 pm Full Moon	10	11	12 Moon near the Pleiades	13 Moon near Aldebaran	14 Moon near Mars	15
16 Moon near Castor and Pollux	17 6.15 pm Last Quarter Moon near Castor and Pollux	18	19	20 Moon near Regulus (am)	21 Orionids	22 Orionids (am)
23	24 Moon near Mercury (am)	25 11.49 am New Moon; partial solar eclipse	26	27	28 Moon near Antares	29

SPECIAL EVENTS

• **5 October:** Saturn lies above the Moon.

• **8 October:** The Moon passes under Jupiter.

• **12–14 October:** The waning Moon passes in turn the Pleiades, Aldebaran and Mars (Chart 10a).

• **Night of 21/22 October:** It's a great year for observing the **Orionid meteor shower,** debris from Halley's Comet smashing into Earth's atmosphere. The shooting stars reach peak performance tonight, and it's deeply dark until the Moon rises around 3.30 am.

• **24 October, before dawn:** Mercury lies below the crescent Moon; best in binoculars (Chart 10b).

• **25 October:** Europe, north-east Africa and the Middle East experience a partial eclipse of the Sun. From the British Isles, the eclipse begins just after 10 am and ends about 11.45 am. At maximum, a few minutes before 11 am, around 15% of the Sun is covered (details depend on your location). DO NOT LOOK DIRECTLY AT THE SUN: see June's Observing Tip for safe methods.

• **30 October, 2 am:** The end of British Summer Time for this year, as clocks go backwards by an hour.

• On **30 and 31 October,** watch out for 'Halloween fireballs', brilliant meteors from Encke's Comet, known as the **Taurids.** They peak next month (see November's Special Events).

10a 12–14 October, 10 pm. The Moon passes the Pleiades and Mars.

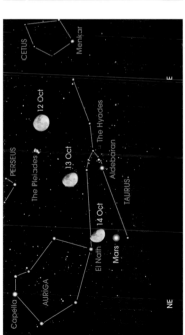

10b 24 October, 7 am. The crescent Moon hangs above Mercury.

- **Saturn** lies in Capricornus. Shining at magnitude +0.6, the ringworld is setting around 1.30 am.
- It's followed by **Neptune**, which sets about 4.30 am. On the border of Aquarius and Pisces, the most distant planet glows at a dim magnitude +7.8.
- Both are far outclassed by brilliant **Jupiter**, at a magnificent magnitude –2.9

it's the brightest object in the night sky, bar the Moon. You'll find the giant planet in Pisces, sinking below the horizon around 5.30 am.

- **Uranus** (magnitude +5.7) lies in Aries, and rises around 7 pm.
- You'll find **Mars** between the horns of Taurus (the Bull), to the left of Aldebaran. The Red Planet comes up above the horizon around 8.30 pm,

and it brightens considerably (from magnitude –0.6 to –1.2) as the Earth draws ever closer.

- If you're up before dawn, watch out for the best morning appearance of **Mercury** this year. The innermost planet lies low on the eastern horizon during the middle of the month, reaching its greatest separation from the Sun on

8 October, when it shines at magnitude –0.4. Mercury will be best seen, however, a week or so later when it brightens to magnitude –1.0.

- **Venus** is lost in the Sun's glare in this month.

OCTOBER'S PLANET WATCH

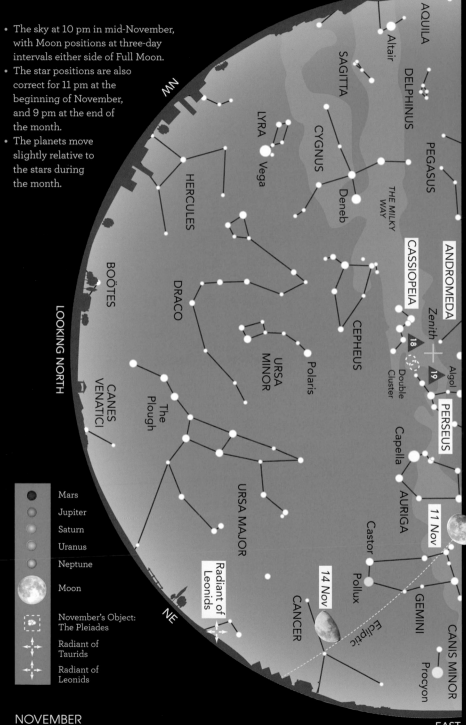

- The sky at 10 pm in mid-November, with Moon positions at three-day intervals either side of Full Moon.
- The star positions are also correct for 11 pm at the beginning of November, and 9 pm at the end of the month.
- The planets move slightly relative to the stars during the month.

WEST

NW

LOOKING NORTH

NE

EAST

AQUILA
Altair
SAGITTA
DELPHINUS
PEGASUS
LYRA
Vega
CYGNUS
Deneb
THE MILKY WAY
ANDROMEDA
CASSIOPEIA
Zenith
HERCULES
18
19
Double Cluster
Algol
BOÖTES
DRACO
CEPHEUS
PERSEUS
URSA MINOR
Polaris
Capella
AURIGA
CANES VENATICI
The Plough
11 Nov
Castor
Pollux
GEMINI
URSA MAJOR
14 Nov
CANCER
CANIS MINOR
Procyon
Ecliptic
Radiant of Leonids

Mars
Jupiter
Saturn
Uranus
Neptune
Moon
November's Object: The Pleiades
Radiant of Taurids
Radiant of Leonids

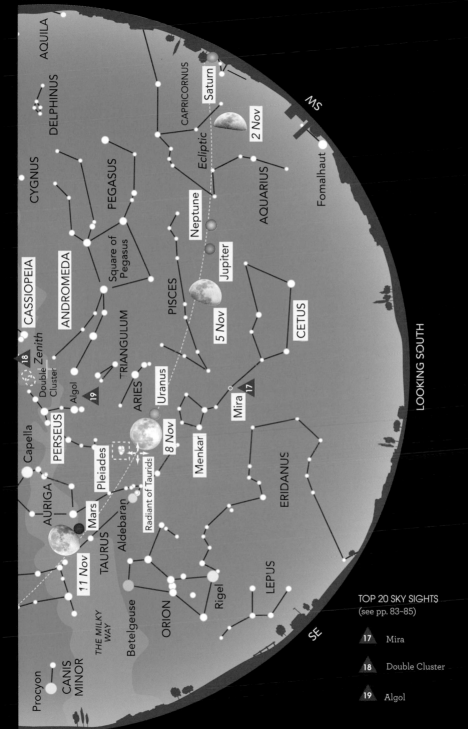

NOVEMBER

AQUILA

DELPHINUS

CYGNUS

CASSIOPEIA

ANDROMEDA

PEGASUS

Square of
Pegasus

Double
Cluster

Zenith

18

CAPRICORNUS

Saturn

Ecliptic

2 Nov

Neptune

Jupiter

5 Nov

PISCES

TRIANGULUM

Algol

19

ARIES

Uranus

8 Nov

Menkar

Mira

17

CETUS

AQUARIUS

Fomalhaut

MS

LOOKING SOUTH

ERIDANUS

Capella

PERSEUS

Pleiades

Radiant of Taurids

AURIGA

Mars

11 Nov

TAURUS

Aldebaran

Betelgeuse

ORION

Rigel

LEPUS

SE

Procyon

CANIS
MINOR

THE MILKY
WAY

TOP 20 SKY SIGHTS
(see pp. 83–85)

17 Mira

18 Double Cluster

19 Algol

EAST

NOVEMBER   **69**

It's another ace month for shooting stars, with the regular **Leonid shower** on 17 November and the chance of brilliant **Taurids** throughout November. Bright planets light up the evening sky, with the scintillating constellations of winter appearing in the east.

NOVEMBER'S CONSTELLATION

In Greek mythology, **Cetus** (the Sea Monster) was a fearsome creature, sent to devour princess **Andromeda** when her mother **Cassiopeia** boasted she was more beautiful than the sea nymphs. Fortunately, the passing superhero **Perseus** slew the ravening beast.

The large dull constellation, though, hardly lives up to this billing. One highlight is **Menkar**, marking the monster's

Roy Stewart captured this atmospheric shot of Comet C2020 F3 NEOWISE at 2.38 am on 11 July 2020, with a tripod-mounted Sony A7iii camera with a Tamron 28-755-mm f/2.8 lens, set to 75 mm. He took a 4-second exposure at ISO 1000.

mouth. Through a telescope it appears as a pretty orange and blue double, though the stars are actually at different distances.

Another is **Mira** (the Wonderful) spotted in 1596 by the German astronomer David Fabricius. Mira then faded, and brightened again – making it the first variable star to be discovered. A distended red giant, some 400 times wider than the Sun, Mira shrinks and swells over a period of 11 months. As a result, the star changes in brightness from magnitude +2 (similar to Polaris) to borderline binocular visibility at magnitude +10. At the moment, it's fading from a maximum last summer.

NOVEMBER'S OBJECT

The **Pleiades** star cluster is one of the most familiar sky sights. It is lovely seen with the naked eye or through binoculars, and magnificent in a long-exposure image.

Though the cluster is well known as the Seven Sisters, skywatchers typically see any number of stars but seven! Most people can pick out the six brightest stars, while very keen-sighted observers can discern up to 11 members. These are just the most luminous in a group of at least 1000 stars, lying 440 light years away. The brightest stars in the Pleiades are hot and blue, and all the stars are young – around 100 million years old (about 2 per cent of the Sun's age).

NOVEMBER'S TOPIC:
VERA RUBIN

Vera Rubin (1928–2016) was fascinated by the night sky from when she was ten years old. She became a leading professional astronomer, battling a strong prejudice against women scientists in the mid-19th-century USA. She was the first female allowed to observe at the famous Palomar Observatory (which then didn't even have women's washrooms).

Rubin was obsessed with galaxies. Astronomers thought a galaxy would rotate rather like the Solar System, with close-in Mercury pelting along and distant Neptune being the plodder. But when Rubin checked galaxies, she found the outermost stars were moving every bit as fast as those further in.

The galaxy's outer regions must be feeling the gravitational of pull of something more than just the stars we see making up the galaxy. Rubin had shown there was invisible material enveloping

OBSERVING TIP

With the brilliant winter constellations starting to appear, you'll probably want to spend plenty of time outside. But make sure you dress up warmly! Lots of layers are better than just a heavy coat, as they trap more air close to your skin. Heavy-soled boots with two pairs of socks stop the frost creeping up your legs. To complete your observing outfit, a woolly hat cuts down on your body heat escaping through the top of your head. And – alas – no hipflask of whisky. Alcohol constricts the veins, and makes you feel even colder.

every star city – what we now call 'dark matter'.

As a tribute to her pioneering work, a ground-breaking astronomical telescope in Chile has been named the Vera C. Rubin Observatory (and it's not short of women's washrooms!).

NOVEMBER'S PICTURE

In July 2020, people in northern latitudes like the British Isles were treated to the brightest comet in over two decades, since Comet Hale-Bopp in 1997. Comet NEOWISE was visible to the naked eye at magnitude +3, hanging low to the north, and a glorious sight in binoculars.

Irish astronomer Roy Stewart set out to observe the comet from one of his favourite dark-sky sites, in the rocky region of the Burren in County Clare. The Poulnabrone dolmen is an ancient burial chamber, over 5000 years old. Adding to the spectacle was a display of noctilucent clouds, a layer of ice crystals right at the top of our atmosphere that can appear on summer nights.

SUNDAY	MONDAY	TUESDAY	WEDNESDAY	THURSDAY	FRIDAY	SATURDAY
		1 6:37 am First Quarter Moon near Saturn	2	3	4 Moon near Jupiter	5
6	7	8 11:02 am Full Moon; lunar eclipse	9 Uranus opposition; Moon between the Pleiades and Aldebaran	10 Moon near Mars	11 Moon near Mars	12
13 Moon near Castor and Pollux	14	15	16 1:27 pm Last Quarter Moon near Regulus	17 Leonid meteor shower	18 Leonid meteor shower (am)	19
20	21 Moon near Spica (am)	22	23 10:57 pm New Moon	24	25	26
27	28	29 Moon near Saturn	30 2:36 pm First Quarter Moon			

★ SPECIAL EVENTS

• Watch out for rare celestial fireworks **throughout November**. Every year, we encounter sparse dust from Encke's Comet, burning up as the feeble **Taurid meteor shower**. But in 2022 Earth passes through a denser region of debris, and pebble-sized rocks may flare across the sky as brilliant fireballs.

• **1 November:** Saturn lies above the Moon (Chart 11a).

• **4 November:** The brilliant 'star' above the Moon is Jupiter (Chart 11a).

• **8 November:** A total eclipse of the Moon is visible from North America and the Pacific, but it starts just after the Moon has set in the British Isles.

• **9 November:** Uranus is opposite to the Sun in the sky and at its closest to Earth, 2796 million kilometres away (see Planet Watch).

• **9 November:** The Moon passes between the Pleiades and Aldebaran, with Mars to the left (Chart 11b).

• **10 November:** The Moon lies between Aldebaran and Mars, with the Pleiades to the upper right (Chart 11b).

• **11 November:** Mars is the bright object near the Moon (Chart 11b).

• **Night of 17/18 November:** Maximum of the annual **Leonid meteor shower.** Good views of these fragments of Comet Tempel-Tuttle impacting our atmosphere as fast shooting stars, before the Moon rises about midnight.

• **29 November:** Saturn lies to the right of the Moon.

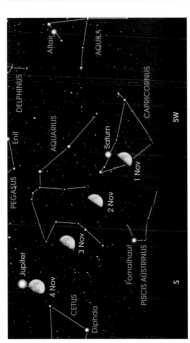

11a 1–4 November, 9 pm. The Moon glides below Saturn and Jupiter.

Saturn

11b 9–11 November, 10 pm. The Moon passes the Pleiades and Mars.

- **Jupiter,** though past its closest to Earth, is still the most brilliant object you'll see in the night sky this month (apart from the Moon). At a glorious magnitude –2.7, the mightiest planet lies in Pisces and is setting around 2 am.

- Just over the border into the next constellation, **Neptune** is a faint magnitude +7.9 in Aquarius. The most distant planet sets about 1.30 am.

- **Saturn** lies well to the lower right of Jupiter, in Capricornus. Setting around 10.30 pm, the ringed planet shines at magnitude +0.7.

- To the left of Jupiter you'll find **Uranus,** above the horizon all night long in Aries. The seventh planet is closest to the Earth on 9 November, but even then it only reaches magnitude +5.6, barely visible to the naked eye. Check it out with

binoculars, which will show Uranus gradually moving from night to night against the starry background.

- At the left-hand end of this planetary procession is **Mars,** rising about 5.30 pm in Taurus. The Red Planet almost doubles in brightness during November, from magnitude –1.2 to –1.8 as the Earth heads towards its close encounter with Mars in December.

- **Mercury** and **Venus** are too close to the Sun to be seen this month.

NOVEMBER'S PLANET WATCH

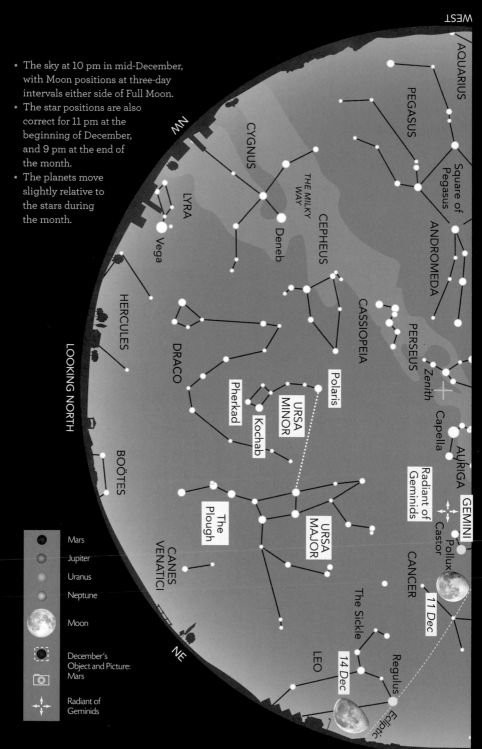

- The sky at 10 pm in mid-December, with Moon positions at three-day intervals either side of Full Moon.
- The star positions are also correct for 11 pm at the beginning of December, and 9 pm at the end of the month.
- The planets move slightly relative to the stars during the month.

WEST

AQUARIUS

PEGASUS

Square of Pegasus

ANDROMEDA

NW

CYGNUS

THE MILKY WAY

CEPHEUS

Deneb

LYRA

Vega

PERSEUS

CASSIOPEIA

Zenith

Capella

HERCULES

DRACO

LOOKING NORTH

Pherkad

Polaris

URSA MINOR

Kochab

AURIGA

Radiant of Geminids

GEMINI

Castor

Pollux

BOÖTES

The Plough

URSA MAJOR

CANCER

11 Dec

Mars

Jupiter

Uranus

Neptune

Moon

December's Object and Picture: Mars

Radiant of Geminids

CANES VENATICI

The Sickle

14 Dec

Regulus

LEO

Ecliptic

NE

EAST

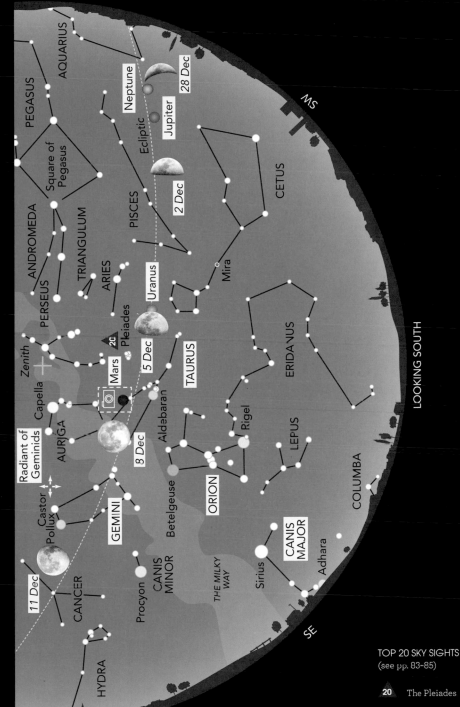

WEST

AQUARIUS

PEGASUS

Square of
Pegasus

PEGASUS

ANDROMEDA

PERSEUS

TRIANGULUM

ARIES

PISCES

Neptune

Ecliptic

Jupiter

28 Dec

2 Dec

MS

CETUS

Uranus

Pleiades

Mira

Zenith

Capella

Mars

5 Dec

20

TAURUS

ERIDANUS

AURIGA

Radiant of
Geminids

8 Dec

Aldebaran

Rigel

LEPUS

Betelgeuse

ORION

COLUMBA

Castor

Pollux

GEMINI

Procyon

CANIS
MINOR

THE MILKY
WAY

CANIS
MAJOR

Sirius

Adhara

11 Dec

CANCER

HYDRA

LOOKING SOUTH

SE

EAST

TOP 20 SKY SIGHTS
(see pp. 83–85)

20   The Pleiades

DECEMBER   75

Mars is this year's 'Christmas Star', at its most brilliant for two years: the Red Planet also takes part in a spectacular party trick when it disappears behind the Full Moon on 8 December. In fact, we have a planetary bonanza this month, with all the planets visible during the evening (use binoculars for **Uranus** and **Neptune**); as well as the seven planets up in the sky, check beneath your feet for the eighth! We are also treated to the sparkling winter constellations **Orion, Canis Major, Taurus** and **Gemini.**

DECEMBER'S CONSTELLATION

Ursa Minor (the Little Bear) is a miniature version of the Great Bear, **Ursa Major,** and it contains perhaps the most famous star: **Polaris,** the Pole Star, which lies almost exactly above Earth's North Pole. Because we spin underneath it, Polaris seems fixed in the sky – due north – and in the past it was an essential reference point for navigators. You can find Polaris by extending a line from the two end stars of the **Plough** (see Star Chart).

The two bears are related, according to legend. The mother was originally a nymph, who had a son by the god Jupiter.

His jealous wife turned the nymph into a bear. Later, Jupiter saw the son about to shoot this apparently wild creature. He quickly gave the young man an ursine shape, too, grabbed both bears by their tails and flung them into the sky.

Polaris (magnitude +2.0, but slightly variable) lies at the tip of the Little Bear's tail. The two stars at the other end of the constellation are known as the 'guardians of the pole': **Kochab** is orange, while **Pherkad** is a blue-white star.

DECEMBER'S OBJECT

Mars – now at its closest to Earth – is the world most similar to our own. The rocky Red Planet has polar caps, seasons, an atmosphere (very thin, and mainly carbon dioxide) and even clouds.

For a planet only half the Earth's size, its geology is truly astonishing. Mars boasts an enormous canyon, Valles Marineris, that is 4000 kilometres long and 7000 metres deep. This enormous crack was caused by the upwelling of the Tharsis Ridge, home to family of vast volcanoes (see Picture).

Largest of the volcanoes – and the biggest in the Solar System – is Olympus Mons. With an altitude of 26 kilometres, the volcano is three times the height of Mount Everest. It would completely cover England, and its central

---

OBSERVING TIP

With Christmas on the way, you may well be thinking of buying a telescope as a present for a budding stargazer. Beware! Unscrupulous websites and mail-order catalogues often advertise small telescopes that boast huge magnifications. This is 'empty magnification' – blowing up an image that the lens or mirror simply doesn't have the ability to get to grips with, so all you see is a bigger blur. The maximum magnification a telescope can provide clearly is twice the diameter of the lens or mirror in millimetres. So if you see an advertisement for a 75-mm telescope, beware of any claims for a magnification greater than 150 times.

crater could swallow London. Although not active today, Olympus Mons may be merely dormant and could erupt again.

Over three dozen successful spacecraft have orbited Mars, landed on its deserts and trundled around to probe its secrets. We know that the Red Planet once had liquid water; before long, scientists should know if life ever flourished on the fourth rock from the Sun.

DECEMBER'S TOPIC: ORIGIN OF LIFE

If there's one question that astronomers are always asked, it's 'How common is life in the Universe?' We have now found over 60 planets like the Earth, orbiting other stars. Living things could thrive there: but has life actually started on those worlds?

The best answer comes from studying how life began on Earth, over 4 billion years ago. The raw materials of life – water and carbon-rich compounds – were abundant. The Earth's own resources were bolstered by prolific supplies delivered by comets and asteroids.

But scientists still argue about where and how these molecules built up into the building blocks of life, and assembled into the first cells. Ancient rock pools, alternately filling up and drying out, could have concentrated the primitive compounds into more complex entities. Or they may have assembled at hydrothermal vents, volcanic fissures on the ocean floor where abundant chemical energy could have kick-started life.

Either way, the creation of living cells took just 200 million years, a mere blink of geological time. Because it happened so rapidly on Earth, many scientists are hopeful that life has evolved on numerous other worlds.

DECEMBER'S PICTURE

This incredibly detailed image isn't from a spacecraft in close proximity to Mars, but was taken from our planet, by the talented planetary photographer Damian Peach when the Red Planet was last close to the Earth two years ago. Based in Selsey on England's south coast, Damian observed Mars remotely with a telescope high in the Andes mountains of Chile (more details on page 88).

The crack spreading left to right across the Red Planet is the great canyon system Valles Marineris; to its right, at the edge of Mars, you can make out three of the planet's great volcanoes. The white spot near the top is Mars's south polar cap (the telescope inverts the image), made of ice and frozen carbon dioxide. Only a few weeks past midsummer, the ice cap is at its minimum size.

On 31 October 2020, Damian Peach pointed the 1-m f/8 Ritchey–Chrétien reflector at the Chilescope Observatory towards the Red Planet, and captured this view with a ZWO ASI174MM camera through red, green and blue filters.

DECEMBER'S CALENDAR

SUNDAY	MONDAY	TUESDAY	WEDNESDAY	THURSDAY	FRIDAY	SATURDAY
				1 Mars closest to Earth; Moon near Jupiter	2	3
4	5 Moon occults Uranus	6 Moon near the Pleiades	7 Moon near Mars and Aldebaran	8 4.08 am Full Moon; Mars opposition; Mars occultation (am)	9	10 Moon near Castor and Pollux
11 Moon near Castor and Pollux	12	13 Geminids; Moon near Regulus	14 Geminids (am)	15	16 8.56 am Last Quarter Moon	17
18 Moon near Spica (am)	19 Moon near Spica (am)	20	21 Winter Solstice; Mercury E elongation	22	23 10.17 am New Moon	24
25 Moon near Venus and Mercury	26 Moon near Saturn	27	28	29 Moon near Jupiter	30 1.20 am First Quarter Moon	31

SPECIAL EVENTS

• **1 December:** Mars is nearer to the Earth than it has been since October 2020, just 81 million km away.

• **1 December:** The Moon lies below giant planet Jupiter.

• **5 December, 4.50–5.20 pm:** The Moon moves in front of Uranus (the exact time depends on your location): view with binoculars or a telescope. This is the second Uranus occultation of 2022 (see September's Special Events).

• **6 December:** The Moon lies just below the Pleiades.

• **7 December:** Mars is right next to the Moon.

• **8 December, 4.55–5.55 am:** Mars suffers a rare occultation by the Full Moon, just minutes after reaching opposition (see Planet Watch and Chart 12a).

• **Night of 13/14 December:** Maximum of the **Geminid meteor shower**, caused by debris from asteroid Phaethon. This year the show is spoilt by the Moon rising around 9 pm.

• **21 December, 9.48 pm:** The Winter Solstice, when the Sun reaches its lowest point in the heavens as seen from the northern hemisphere, giving us the shortest day and the longest night.

• **25 December:** To the lower right of the crescent Moon is Venus, with fainter Mercury (best seen in binoculars) lying between them (Chart 12b).

• **26 December:** Saturn lies above the crescent Moon.

• **29 December:** Jupiter lies just above the Moon.

Mars, disappearance

Mars, reappearance

12a 8 December, 4.55–5.55 am. The Full Moon occults Mars.

Saturn

CAPRICORNUS

AQUILA

SAGITTARIUS

Mercury

Venus

SSW

SW

12b 25 December, 4.45 pm. The crescent Moon near Venus and Mercury.

- This month, **Venus** begins another long reign as the Evening Star, appearing low in the south-west just after sunset. By the end of December, Venus blazes at magnitude –3.9 and sets around 5.30 pm.
- For most of the month, you'll find **Mercury** to the upper left of Venus, and 20 times fainter (best seen in binoculars). At its greatest separation from the Sun on 21 December, the innermost

planet shines at magnitude –0.5 and sets about 5 pm.
- **Saturn**, at magnitude +0.8 embellishes the faint stars of Capricornus and sets around 8.30 pm.
- It's followed by dim **Neptune** (magnitude +7.9), setting about 11.30 pm in Aquarius.
- Mighty **Jupiter** is the brightest planet after Venus has set, at magnitude –2.5 in Pisces. It sinks below the horizon just after midnight.

- **Uranus** (magnitude +5.7) lies in Aries and sets about 4.30 am. On 5 December, it's hidden by the Moon (see Special Events).
- **Mars** is visible all night long, at its closest to Earth on 1 December, lying opposite the Sun a week later, and reaching a maximum brilliance of magnitude –1.9 (almost rivalling Jupiter). But there's even more in store!
- On the morning of 8 December, Mars reaches

opposition at 4.24 am, just 16 minutes after the Full Moon, so the Sun, the Earth, the Moon and Mars are almost exactly in line. Then the Moon moves right in front of Mars, in another rare planetary occultation. Mars disappears about 4.55 am and re-emerges around 5.55 am (the exact time varies depending on your location). It's the first Mars occultation visible from the British Isles since 1952; the next won't occur until 2052.

Can you see the planets? It's a common question; and the answer is a resounding 'yes!' Some of our cosmic neighbours are the brightest objects in the night sky after the Moon. As they're so close, you can watch them getting up to their antics from night to night. And planetary debris – leftovers from the birth of the Solar System – can light up our skies as glowing comets and the celestial fireworks of a meteor shower.

THE SUN-HUGGERS

Mercury and Venus orbit the Sun more closely than our own planet, so they never seem to stray far from our local star: you can spot them in the west after sunset, or the east before dawn, but never all night long. At *elongation*, the planet is at its greatest separation from the Sun, though – as you can see in the diagram (right) – that's not when the planet is at its brightest. Through a telescope, Mercury and Venus (technically known as the *inferior planets*) show phases like the Moon – from a thin crescent to a full globe – as they orbit the Sun.

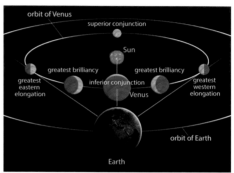

Venus (and Mercury) show phases like the Moon as they orbit the Sun.

Mercury

The innermost planet appears in the evening sky in January, makes its best appearance in April–May, and then reappears in the dusk sky in December (its evening appearance in July–August is lost in the bright twilight). Mercury is low in the dawn twilight at its February–March and June apparitions; it's best seen before dawn in October.

Venus

Venus bookends 2022 as the Evening Star, during the first few days of January and again in December. Otherwise, it's in the morning sky all year, most prominent in February, when the Morning Star rises from the dawn glow into a darker sky.

Maximum elongations of Mercury in 2022	
Date	Separation
7 January	19° east
16 February	26° west
29 April	21° east
16 June	23° west
27 August	27° east
8 October	18° west
21 December	20° east

Maximum elongation of Venus in 2022	
Date	Separation
20 March	47° west

WORLDS BEYOND

A planet orbiting the Sun beyond the Earth (known in the jargon as a *superior planet*) is visible at all times of night, as we look outwards into the Solar System. It lies due south at midnight when the Sun, the Earth and the planet are all in line – a time known as *opposition* (see the diagram, right). Around this time the Earth lies nearest to the planet, although the date of closest approach (and the planet's maximum brightness) may differ by a few days because the planets' orbits are not circular.

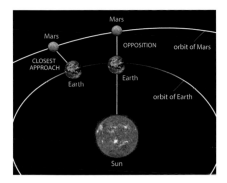

Mars (and the outer planets) line up with the Sun and Earth at opposition, but they are brightest at their point of closest approach.

Mars

For the first half of the year, you'll need to be up after midnight to catch the Red Planet. In the autumn, Mars brightens rapidly towards closest approach on **1 December** and opposition on **8 December**, when also it undergoes a rare occultation by the Moon.

● Where to find Mars	
Early January	Ophiuchus
Late January–February	Sagittarius
March–early April	Capricornus
Late April–early May	Aquarius
Late May–June	Pisces
July–early August	Aries
Late August–December	Taurus

Jupiter

The giant planet starts the year low in the evening sky in Aquarius, sinking into the sunset by mid-February. Jupiter reappears in the morning sky at the beginning of April, in Pisces where it remains for the rest of the year (except in July and August when it strays slightly from the traditional Zodiac constellations into neighbouring Cetus). Reaching opposition on **26 September**, Jupiter is visible in the evening sky until the end of 2022.

Saturn

During the first half of January, you can catch the ringed planet just above the horizon after sunset. After disappearing into the twilight glow, Saturn reappears before dawn towards the end of February. Saturn is at opposition on **14 August**, and you'll then see it in the evening sky until the close of 2022. It resides in Capricornus all year.

Uranus

Just perceptible to the naked eye, Uranus lies in Aries all year. Up until May, the seventh planet is visible in the evening sky. It emerges from the Sun's glow in the morning sky in June. Uranus is at opposition on **9 November**. The planet suffers rare occultations by the Moon on **14 September** and **5 December**.

Neptune

The most distant planet lies on the border of Aquarius and Pisces throughout the year, and is at opposition on **16 September**. Neptune can be seen (though only through binoculars or a telescope) in January and February and then from the end of April until the end of the year.

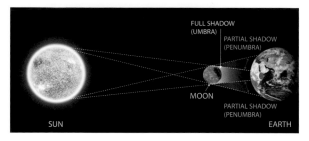

Where the dark central part (the umbra) of the Moon's shadow reaches the Earth, we are treated to a total solar eclipse. If the shadow doesn't quite reach the ground, we see an annular eclipse. People located within the penumbra observe a partial eclipse.

SOLAR ECLIPSES

On **30 April**, south-western South America and adjacent regions of the Pacific Ocean will experience a partial eclipse of the Sun; it's not total anywhere on Earth.

On **25 October** there's a partial solar eclipse (not total anywhere) visible from Europe, north-east Africa and the Middle East. From the British Isles, the eclipse begins just after 10 am and ends about 11.45 am. Maximum obscuration occurs a few minutes before 11 am, when about 15 per cent of the Sun is covered.

LUNAR ECLIPSES

A total lunar eclipse on **16 May** is visible from the Americas and western regions of Africa and Europe. From the British Isles, the eclipse can be seen low in the south-west in the early morning sky, with the Moon setting during totality

On **8 November**, anyone in North America and the Pacific will be treated to a total eclipse of the Moon; but nothing is visible from the British Isles.

METEOR SHOWERS

Shooting stars, or *meteors,* are tiny specks of interplanetary dust, burning up in the Earth's atmosphere. At certain times of year, Earth passes through a stream of debris (usually left by a comet) and we see a *meteor shower.* The meteors appear to emanate from a point in the sky known as the *radiant.* Most showers

Table of annual meteor showers	
Meteor shower	Date of maximum
Quadrantids	3/4 January
Lyrids	21/22 April
Eta Aquarids	6 May (am)
Perseids	12/13 August
Orionids	21/22 October
Leonids	17/18 November
Geminids	13/14 December

are known by the constellation in which the radiant lies.

In addition to the regular showers, this year we may be treated to spectacular displays on **30 May/1 June** (the Tau Herculids) and from the **end of October throughout November** (the Taurids).

COMETS

Comets are dirty snowballs from the outer Solar System. If they fall towards the Sun, its heat evaporates their ices to produce a gaseous head (*coma*) and sometimes dramatic tails. Although some comets are visible to the naked eye, use binoculars to reveal stunning details in the coma and the tail.

Hundreds of comets move round the Sun in small orbits. But many more don't return for thousands or even millions of years. Most comets are now discovered in professional surveys of the sky, but a few are still found by dedicated amateur astronomers. Watch out in case a brilliant new comet puts in a surprise appearance!

Here are some of the most popular sights in the night sky, in a season-by-season summary. It doesn't matter if you're a complete beginner, finding your way around the heavens with the unaided eye ◉ or binoculars 🔭; or if you're a seasoned stargazer, with a moderate telescope ⟩. There's something here for everyone.

Each sky sight comes with a brief description, and a guide as to how you can best see it. Many of the most delectable objects are faint, so avoid moonlight when you go out spotting. Most of all, enjoy!

SPRING

Praesepe ◉ 🔭 ⟩

Constellation: Cancer
Star Chart/Key: March;
Type/Distance: Star cluster; 600 light years
Magnitude: +3.7
A fuzzy patch to the unaided eye; a telescope reveals many of its 1000 stars.

M81 and M82 🔭 ⟩

Constellation: Ursa Major
Star Chart/Key: March; 6
Type/Distance: Galaxies; 12 million light years
Magnitude: +6.9 (M81); +8.4 (M82)
A pair of interacting galaxies: the spiral M81 appears as an oval blur, and the starburst M82 as a streak of light.

The Plough

The Plough ◉

Constellation: Ursa Major
Star Chart/Key: April;
Type/Distance: Asterism; 82–123 light years
Magnitude: Stars are roughly magnitude +2
The seven brightest stars of the Great Bear form a large saucepan shape, called 'the Plough'.

Mizar and Alcor ◉ 🔭 ⟩

Constellation: Ursa Major
Star Chart/Key: April; 8
Type/Distance: Double star; 83 & 82 light years
Magnitude: +2.3 (Mizar); +4.0 (Alcor)
The sky's classic double star, easily separated by the unaided eye: a telescope reveals Mizar itself is a close double.

Virgo Cluster 🔭 (difficult) ⟩

Constellation: Virgo
Star Chart/Key: May; 9
Type/Distance: Galaxy cluster; 54 million light years
Magnitude: Galaxies range from magnitude +9.4 downwards
Huge cluster of 2000 galaxies, best seen through moderate to large telescopes.

SUMMER

Antares ◉ 🔭 ⟩

Constellation: Scorpius
Star Chart/Key: June; 10
Type/Distance: Double star; 600 light years
Magnitude: +0.96
Bright red star close to the horizon. You can spot a faint green companion with a telescope.

M13

Constellation: Hercules
Star Chart/Key: June;
Type/Distance: Star cluster; 22,200 light years
Magnitude: +5.8
A faint blur to the naked eye, this ancient globular cluster is a delight seen through binoculars or a telescope. It boasts nearly a million stars.

Lagoon and Trifid Nebulae

Constellation: Sagittarius
Star Chart/Key: July; 12
Type/Distance: Nebulae; 5000 light years
Magnitude: +6.0 (Lagoon); +7.0 (Trifid)
While the Lagoon Nebula is just visible to the unaided eye, you'll need binoculars or a telescope to spot the Trifid. The two are in the same binocular field of view, and present a stunning photo opportunity.

Albireo

Constellation: Cygnus
Star Chart/Key: August; 13
Type/Distance: Double star; 430 light years
Magnitude: +3.2 (Albireo A) ; +5.1 (Albireo B)
Good binoculars reveal Albireo as being double. But you'll need a small telescope to appreciate its full glory. The brighter star appears golden; its companion shines piercing sapphire. It is the most beautiful double star in the sky.

Dumbbell Nebula

Constellation: Vulpecula
Star Chart/Key: August; 14
Type/Distance: Planetary nebula; 1300 light years
Magnitude: +7.5

Dumbbell Nebula

Andromeda Galaxy

Visible through binoculars, and a lovely sight through a small/medium telescope, this is a dying star that has puffed off its atmosphere into space.

AUTUMN

Delta Cephei

Constellation: Cepheus
Star Chart/Key: September; 15
Type/Distance: Variable star; 890 light years
Magnitude: +3.5 to +4.4, varying over 5 days 9 hours
The classic variable star, Delta Cephei is chief of the Cepheids – stars that allow us to measure distances in the Universe (their variability time is coupled to their intrinsic luminosity). Visible to the unaided eye, but you'll need binoculars for serious observations.

Andromeda Galaxy

Constellation: Andromeda
Star Chart/Key: October; 16
Type/Distance: Galaxy; 2.5 million light years
Magnitude: +3.4
The nearest major galaxy to our own, the Andromeda Galaxy is easily visible to the unaided eye in unpolluted skies. Four times the width of the Full Moon, it's a great telescopic object and photographic target.

Mira

Constellation: Cetus
Star Chart/Key: November; 17
Type/Distance: Variable star; 300 light years
Magnitude: +2 to +10 over 332 days, although maxima and minima may vary.

Nicknamed 'the Wonderful', this distended red giant star is alarmingly variable as it swells and shrinks. At its brightest, Mira is a naked-eye object; binoculars may catch it at minimum; but you need a telescope to monitor this star. Its behaviour is unpredictable, and it's important to keep logging it.

Double Cluster ◉ 🏋 ↘

Constellation: Perseus
Star Chart/Key: November;
Type/Distance: Star clusters; 7500 light years
Magnitude: +3.7 and +3.8
A lovely sight to the unaided eye, these stunning young star clusters are sensational through binoculars or a small telescope. They're a great photographic target.

Algol ◉ 🏋 ↘

Constellation: Perseus
Star Chart/Key: November;
Type/Distance: Variable star; 90 light years
Magnitude: +2.1 to +3.4 over 2 days 21 hours
Like Mira, Algol is a variable star, but not an intrinsic one. It's an 'eclipsing binary' – its brightness falls when a fainter companion star periodically passes in front of the main star. It's easily monitored by the eye, binoculars or a telescope.

WINTER
Pleiades ◉ 🏋 ↘

Constellation: Taurus
Star Chart/Key: December;
Type/Distance: Star cluster; 440 light years
Magnitude: Stars range from magnitude +2.9 downwards
To the naked eye, most people can see six stars in the cluster, but it can rise to 14 for the keen-sighted. In binoculars or a telescope, they are a must-see. Astronomers have observed 1000 stars in the Pleiades.

Orion Nebula ◉ 🏋 ↘

Constellation: Orion
Star Chart/Key: January;
Type/Distance: Nebula; 1340 light years
Magnitude: +4.0

A striking sight even to the unaided eye, the Orion Nebula – a star-forming region 24 light years across – hangs just below Orion's belt. Through binoculars or a small telescope, it is staggering. A photographic must!

Betelgeuse ◉ 🏋

Constellation: Orion
Star Chart/Key: January;
Type/Distance: Variable star; 720 light years
Magnitude: 0.0 to +1.3
Even with the unaided eye, you can see that Betelgeuse is slightly variable over months, as the red giant star billows in and out.

M35 ◉ 🏋 ↘

Constellation: Gemini
Star Chart/Key: February;
Type/Distance: Star cluster; 2800 light years
Magnitude: +5.3
Just visible to the unaided eye, this cluster of around 2000 stars is a lovely sight through a small telescope.

Sirius ◉ 🏋 ↘

Constellation: Canis Major
Star Chart/Key: February;
Type/Distance: Double star; 8.6 light years
Magnitude: –1.47
You can't miss the Dog Star. It's the brightest star in the sky! But you'll need a 150-mm reflecting telescope (preferably bigger) to pick out its +8.44 magnitude companion – a white dwarf nicknamed 'the Pup'.

Pleiades

Photos of the night sky, such as those in this book, enchant us – and may leave us wondering how such amazing photos were taken. So here are some of the stories behind the pictures, to show how they were achieved, and maybe even encourage you to attain similar results.

They were taken with a wide range of equipment, from a simple holiday grab shot using a phone camera to a large mountaintop telescope with professional facilities. But none were taken by professional astronomers with exclusive access, and some didn't require any particularly expensive outlay or access to dark skies.

STAR CAMERAS

These days, even a phone camera can take dramatic sky photos. Although Nigel Henbest's photo of the Moon, Jupiter and Saturn (page 53) was taken from Barbados using a Samsung Galaxy S10e, it could have been taken from anywhere with a clear sky and a nice foreground. Nor are phone cameras limited to shots at dusk. With their night settings, or using software that allows longer time exposures, recent phones can easily take photos of constellations and even aurorae. In terms of cost, given that you probably have the phone anyway, these provide the cheapest astro shots you can take.

But you do need to keep the phone rock steady during the time exposure. You can get inexpensive stands that won't take up much space in your luggage and will make all the difference to the shot, or you can improvise.

Next up are everyday cameras, particularly DSLRs. They are more versatile than phone cameras, allowing a greater range of time exposures and lens changes. But get to know how to use them before you see that great photo opportunity – you

need to master the manual settings, and of course a tripod is vital. Again, if you have the camera anyway, the additional outlay is fairly low cost, either just a tripod or a relatively inexpensive portable tracking mount that counteracts the diurnal movement of the heavens, allowing time exposures with telephoto lenses.

Then there are the dedicated astro-cameras. These are designed to be linked to computers rather than being stand-alone devices. Many are mono (black and white) only, and are used with specialised filters to give a final colour result. These are generally used on telescopes on hefty mountings that track the sky very precisely, with autoguiders to correct any errors in the system. Most of these involve a considerable outlay, both financially and in time and expertise. And just taking the exposures is only part of the effort – there's often a lot of processing needed to get the most out of every image.

These home-grown systems are now joined by remote observing, in which the equipment is not yours but in a dark-sky overseas site and rented by the minute on a monthly points basis. The cost per minute may initially seem high, but even at the starter rate, for the cost of about four cups of coffee-shop coffee you can typically take around 10 minutes of exposures on a 250-mm telescope in black skies, ready to go and set up – or 4 minutes on a half-metre telescope. And if clouds roll over or something goes wrong, you are not charged.

Finally, you can process data obtained by space probes such as Juno at fabulous expense, but made available free of charge. So, paradoxically, this is the cheapest form of astrophotography you can get!

THE PHOTOS

Jo Bourne's star trail photo from Cissbury Ring in Sussex (page 23) is an example of a spectacular picture that can be taken using basic equipment, a tripod and some patience from anywhere with a reasonably dark sky. Her camera is a Canon 60D, but you could use any camera with manual exposure control that allows 30-second exposures. Pointing her camera at the Pole Star, she used a wide-angle lens and chose the setting of ISO 400 to give well-exposed stars. If the ISO is too low you will get too few stars in a 30-second exposure, and if it's too high there could be too much background light even in a fairly dark area such as the South Downs. For about an hour she shot one exposure after another using a cable release set for continuous exposures, taking a total of 114.

Then she used the free stacking software StarStaX to add all the images together. This includes a nifty option to fill in the inevitable short gaps between the individual shots. Finally, she adjusted the colour and brightness settings using the paid-for software Adobe Lightroom, although free software such as GIMP would also do the job.

Nigel Bradbury's photo of the aurora on page 11 was taken using similar equipment, although in this case he was definitely in the right place at the right time. 'With all Northern Lights tours cancelled in Iceland due to stormy weather,

Jo Bourne used this camera and tripod to achieve her star trail photo. The anti-dew heater band around the lens barrel is a useful addition in the British climate.

Reykjavik was soon devoid of anybody outside at night, with rain coming in sideways!' Nigel reports. 'But I chose to stay out all night. The clouds blew a tad clearer by 4 am, allowing this shot as I stood on the waterfront. I took great delight in showing those at breakfast what they had just missed outside!'

A step up from the basic shots using just a camera on a tripod is shown in the photo on the title page of Comet NEO-WISE by Damian Peach and Ian Sharp. As they used a 200-mm focal length lens, the stars would have started to trail after just a few seconds' exposure with a fixed camera, so in order to build up a good image they needed to use a telescope tracking mount to give 51 × 15-second exposures. However, even though the comet was 125 million kilometres away, its movement through space against the starry background was evident between individual exposures. So they needed to stack the frames separately on the comet and the stars, and carefully merge the two stacks to give the final result.

The impressive array of Chilescope domes in the dark skies of Chile are among the increasing range of instruments available for remote observing.

Although Damian often uses his own 355-mm telescope from his home near Southampton, the planets don't always rise high enough from the UK to give the best views. So for his photo of the conjunction of Jupiter and Saturn on page 17, which took place with the planets only a few degrees above his horizon on 21 December 2020, he used the 1-metre Chilescope, a remote share instrument located in the Atacama desert, just 20 kilometres from the professional Gemini South Observatory and in an area offering both steady viewing and dark skies. This instrument has both planetary and deep-sky cameras. Furthermore, the planets were 18 degrees above the horizon. Damian used the same instrument for his picture of Mars on page 77.

Remote observing gets the results, but for many the challenge of using your own equipment is very worthwhile. Simon Hudson's very deep image of the galaxies M81 and M82 on page 65 required both a comparatively wide-angle field of view and long exposure times. You might imagine that a photo of galaxies would need a big telescope, but Simon used a 70-mm refractor – the same aperture as many beginners' telescopes. The difference is that his Altair Astro 70 EDQ-R has a carefully designed four-element lens to overcome the false colour inherent in basisc refractors, and is effectively a 350-mm focal length f/5 telephoto lens.

And to get the surrounding integrated flux nebulosity – the cold gas and dust between the stars in our own Galaxy – Simon needed to give very long exposure times from his home in Cornwall to the north of Truro. He used 76 × 10-minute unfiltered shots on his mono astro-camera to record the faintest detail, then 24 × 5-minute exposures through each red, green and blue filter to add colour. 'This was around 25 hours of data, so the most I've spent on an image,' says Simon.

This required gathering data over several nights, but as Simon doesn't have an observatory he needed to set up his equipment each night, realigning the telescope mount on the north celestial pole and reacquiring exactly the same field of view. Many of these tasks are assisted using software such as Sequence Generator Pro and SharpCap.

Sara Wager, on the other hand, has an observatory at her home in Spain so her equipment is ready to go each night. In May 2020 she polar-aligned her telescope for the first time in two years! Her photo of the Needle Galaxy on page 35 required over 37 hours of exposures so for many observers in British weather that would occupy a whole season! But Sara says, 'To acquire a target night to night is as simple as opening the roof, switching on the PC and software, connecting all the hardware then plate solving the previous image and then I am within six pixels of the sample image and ready to image for another night. On average I get 7–8 hours of data per night.' Plate solving software compares the stars in an image with a star catalogue and works out exactly which part of the sky has been photographed.

Pete Williamson uses equipment that's out of this world – quite literally. His image of Jupiter on page 59 was made using NASA's Juno spacecraft, currently orbiting the giant planet. However, NASA scientists don't examine in detail all the images from Juno, as it sweeps closely over Jupiter every 53 days. They rely on amateur enthusiasts to help process and analyse its views of the planet's

In Sara Wager's observatory the telescope and MESU200 mount are on a solid pillar, which greatly improves stability. The filter wheel assembly, between the CCD camera (blue) and the telescope, includes an off-axis guider that uses a guide star outside the main field of view, rather than a separate guidescope.

ever-changing clouds, and to upload the resulting images to their website.

During his regular perusal of the Juno-Cam data library (missionjuno.swri.edu/junocam), Pete picked out this unfamiliar angle, unavailable from the viewpoint of Earth. He downloaded NASA's simple bland image, and carefully processed it to bring out details in the colours of the planet's intricate cloud system. Pete also uses data from the 2-metre Faulkes Telescopes in Hawaii and Australia, which are available free of charge to education and official outreach users.

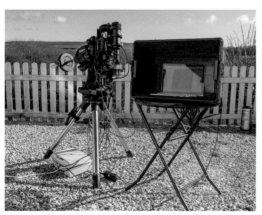

Although Simon Hudson's 70-mm refractor has a small aperture, the setup needed for long-exposure imaging includes a substantial equatorial mount, a separate guide telescope and camera on top of the main instrument, dew heaters and telescope control systems, all run from a laptop.

Our view of the stars – a source of infinite amazement for scientists, stargazers and the millions of us who seek out rural places to rest and recuperate – is obscured by light pollution. It's a sad fact that many people may never see the Milky Way, our own Galaxy, because of the impact of artificial light.

LIGHT POLLUTION

Light pollution is a generic term referring to artificial light that shines where it is neither wanted nor needed. In broad terms, there are three types of light pollution:

- **Skyglow** – the pink or orange glow we see for miles around towns and cities, spreading deep into the countryside, caused by a scattering of artificial light by airborne dust and water droplets.
- **Glare** – the uncomfortable brightness of a light source.
- **Light intrusion** – light spilling beyond the boundary of the property on which a light is located, sometimes shining through windows and curtains.

CPRE, the countryside charity, has long fought for the protection and improvement of dark skies, and against the spread of unnecessary artificial light. CPRE commissioned LUC to create new maps of Great Britain's light pollution and dark skies to give an accurate picture of how much light is spilling up into the night sky and show where urgent action is needed. CPRE also sought to find where the darkest skies are, so that they can be protected and improved.

MAPPING

The maps are based on data gathered by the National Oceanographic and Atmospheric Administration (NOAA) in America, using the Suomi NPP weather satellite. One of the instruments on board the satellite is the Visible Infrared Imaging Radiometer Suite (VIIRS), which captures visible and infrared imagery to monitor and measure processes on Earth, including the amount of light spilling up into the night sky. This light is captured by a day/night band sensor.

The mapping used data gathered in September 2015, and is made up of a composite of nightly images taken that month as the satellite passed over the UK at 1.30 am.

The data was split into nine categories to distinguish between different light levels. Colours were assigned to each category, ranging from darkest to brightest, as shown in the chart below. The maps

Colour bandings to show levels of brightness

Categories	Brightness values (in nw/cm²/sr)*
Colour band 1 (darkest)	<0.25
Colour band 2	0.25–0.5
Colour band 3	0.5–1
Colour band 4	1–2
Colour band 5	2–4
Colour band 6	4–8
Colour band 7	8–16
Colour band 8	16–32
Colour band 9 (brightest)	>32

The brightness values are measured in nano-watts/cm²/steradian (nw/cm²/sr). In simple terms, this calculates how the satellite instruments measure the light on the ground, taking account of the distance between the two.

are divided into pixels, 400 metres × 400 metres, to show the amount of light shining up into the night sky from that area. This is measured by the satellite in nanowatts, which is then used to create a measure of night-time brightness.

The nine colour bands were applied to a national map of Great Britain (see the following pages), which clearly identifies the main concentrations of night-time lights, creating light pollution that spills up into the sky.

The highest levels of light pollution are around towns and cities, with the highest densities around London, Leeds, Manchester, Liverpool, Birmingham and Newcastle. Heavily lit transport infrastructure, such as major roads, ports and airports, also show up clearly on the map. The national map also shows that there are many areas that have very little light pollution, where people can expect to see a truly dark night sky.

The results show that only 21.7 per cent of England has pristine night skies, completely free from light pollution (see the chart below). This compares with almost 57 per cent of Wales and 77 per cent of Scotland. When the two darkest categories are combined, 49 per cent of England can be considered dark, compared with almost 75 per cent in Wales and 87.5 per cent in Scotland. There are noticeably higher levels of light pollution in England in all the categories, compared with Wales and Scotland. The amount of the most severe light pollution is five times higher in England than in Scotland and six times higher than in Wales.

The different levels of light pollution are linked to the varying population densities of the three countries: where there are higher population densities, there are higher levels of light pollution. For example, the Welsh Valleys are clearly shown by the fingers of light pollution spreading north from Newport, Cardiff, Bridgend and Swansea. In Scotland, the main populated areas stretching from Edinburgh to Glasgow show almost unbroken levels of light pollution, creeping out from the cities and towns to blur any distinction between urban and rural areas.

Light levels in England, Wales and Scotland

Categories	England	Wales	Scotland	GB
Colour band 1 (darkest)	21.7%	56.9%	76.8%	46.2%
Colour band 2	27.3%	18.0%	10.7%	20.1%
Colour band 3	19.1%	9.3%	4.6%	12.6%
Colour band 4	11.0%	5.8%	2.8%	7.3%
Colour band 5	6.8%	3.8%	1.7%	4.6%
Colour band 6	5.0%	2.9%	1.2%	3.3%
Colour band 7	4.3%	2.1%	1.0%	2.8%
Colour band 8	3.2%	1.0%	0.9%	2.1%
Colour band 9 (brightest)	1.6%	0.2%	0.3%	1.0%

Adapted from Night Blight: Mapping England's light pollution and dark skies *CPRE (2016), with kind permission from CPRE. To see the full report and dedicated website, go to http://nightblight.cpre.org.uk/*

MAP OF BRITAIN'S LIGHT POLLUTION AND DARK SKIES
COURTESY OF CPRE/LUC

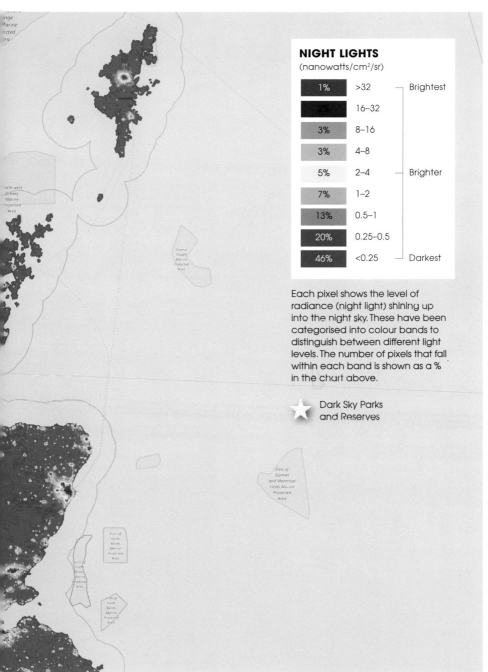

NIGHT LIGHTS

(nanowatts/cm²/sr)

	%	Value	Band
■	1%	>32	Brightest
■	2%	16–32	
■	3%	8–16	
■	3%	4–8	
■	5%	2–4	Brighter
■	7%	1–2	
■	13%	0.5–1	
■	20%	0.25–0.5	
■	46%	<0.25	Darkest

Each pixel shows the level of radiance (night light) shining up into the night sky. These have been categorised into colour bands to distinguish between different light levels. The number of pixels that fall within each band is shown as a % in the chart above.

★ Dark Sky Parks and Reserves

© OpenStreetMap contributors, Earth Observation Group, NOAA National Geophysical Data Center. Developed by CPRE and LUC.

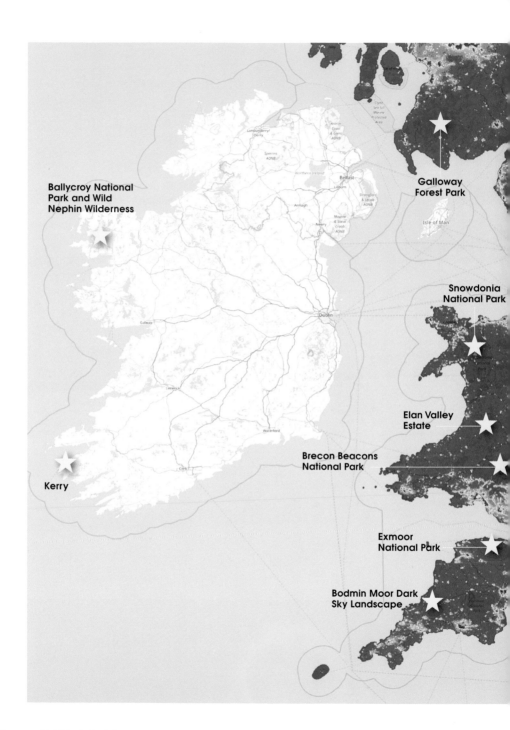

Ballycroy National
Park and Wild
Nephin Wilderness

Galloway
Forest Park

Snowdonia
National Park

Elan Valley
Estate

Brecon Beacons
National Park

Kerry

Exmoor
National Park

Bodmin Moor Dark
Sky Landscape

Moffat

Northumberland National
Park and Kielder Water
and Forest Park

Dark Sky Parks
and Reserves

North York Moors National Park

Yorkshire Dales National Park

Moore's Reserve
(South Downs)

Cranborne Chase
Area of
Outstanding
Natural Beauty

THE AUTHOR

Nigel Henbest is an award-winning British science writer, specialising in astronomy and space. After research in radio astronomy at the University of Cambridge, he became a consultant to both the Royal Greenwich Observatory and *New Scientist* magazine, and is a Fellow of the Royal Astronomical Society.

With the late Heather Couper, Nigel has been writing Philip's *Stargazing* since 2005. He has penned over 40 other books, along with hundreds of articles in magazines and newspapers. Nigel appears regularly on radio and television to comment on breaking astronomy news. He co-founded a TV production company,

where he has produced acclaimed programmes and international series on astronomy and space.

Nigel is a Future Astronaut with Virgin Galactic, and asteroid 3795 is named 'Nigel' in his honour.

ACKNOWLEDGEMENTS

PHOTOGRAPHS
Front cover: Pete Williamson. **Jo Bourne:** 23, 87. **Nigel Bradbury:** 11. **James Harrop:** 29. **Nick Hart:** 85. **Bridget Henbest:** 96. **Nigel Henbest:** 7, 53. **Peter Jenkins:** 41. **Landscape & Sky by Simon Hudson:** 65, 89 bottom. **Pete Lawrence:** 46. **NASA:**/UCLA 2–3;/JPL-Caltech 18, 24;/Bill Ingalls 48;/University of Arizona 49;/J. Clarke (Boston University) and G. Bacon (STScI) 73;/VegaStar Carpentier 83;/Robert Gendler 84 (top). **Damian Peach:** 1, 17, 77, 88. **Robin Scagell:** 6, 84 (bottom). **Roy Stewart (Skellig Star Party):** 70. **Sara Wager – www.swagastro.com:** 35, 89 top. **Pete Williamson FRAS:** 59.

ARTWORKS
Star maps: Wil Tirion/Philip's with extra annotation by Philip's.
Planet event charts: Nigel Henbest/Stellarium (www.stellarium.org).
pp. 80–82: Chris Bell.
pp. 90–95: Adapted from *Night Blight: Mapping England's light pollution and dark skies* CPRE (2016), with kind permission from CPRE. To see the full report and dedicated website, go to http://nightblight. cpre.org.uk/Maps © OpenStreetMap contributors, Earth Observation Group, NOAA National Geophysical Data Center. Developed by CPRE and LUC.